QUANTUM MECHANICS IN SIMPLE MATRIX FORM

QUANTUM MECHANICS IN SIMPLE MATRIX FORM

THOMAS F. JORDAN

Physics Department
University of Minnesota, Duluth

A Wiley-Interscience Publication

JOHN WILEY & SONS

New York Chichester Brisbane Toronto Singapore

Library of Congress Cataloging in Publication Data:

Jordan, Thomas, F., 1936–
Quantum mechanics in simple matrix form.

"A Wiley-Interscience publication."
Includes index.
1. Quantum theory. 2. Matrices. I. Title.

QC174.12.J67 1985 530.1'2 85-12121
ISBN 0-471-81751-1

Printed in the United States of America

10 9 8 7 6 5 4 3

This book is dedicated
with gratitude and best wishes to

HOWARD G. HANSON

on the occasion of his retirement
after 33 years as
head of the physics department
at the University of Minnesota, Duluth,
where I have learned much
from his teaching and his example.

PREFACE

This book is both very modest and very ambitious. The extent of its subject is modest. It is not a complete book of quantum physics. It treats only the mathematical language of quantum mechanics, the theoretical structure, and that only from the matrix point of view. Other books and sources have to be used along with it to get a full account of either quantum mechanics including wave mechanics or the experiments that reveal the phenomena quantum mechanics describes.

The book is ambitious in making basic quantum mechanics accessible with minimum mathematics. It avoids the mathematics of Hilbert space, Hermitian and unitary matrices, eigenvalues and eigenvectors, and the like. There are no state vectors or wave functions at all. There are no differential equations. The book does not even use calculus or trigonometry. It assumes only basic algebra.

The emphasis is on the matrices representing physical quantities. States are described simply by mean values of physical quantities or equivalently by probabilities for different possible values. This requires using the algebra of matrices and complex numbers together with probabilities and mean values. These bits of mathematics are introduced at the beginning and then used over and over.

I was surprised to find how much can be done this way, with one hand tied. It is not only possible; for many things it is easier or better. For example, calculating correlations of two spins is easier without state vectors. The absence of a continuous range of possible values for angular momentum or oscillator energy comes out more clearly than in standard methods.

Much of this is original. I found new ways to do things. I hope these will be interesting to anyone who teaches quantum mechanics at any level. I use some of them in my graduate course.

This approach reveals the essential simplicity of quantum mechanics by focusing on the bare skeleton and working only with the key elements of the mathematical structure.

The book grew out of a course I teach for college students not majoring in physics. I also use it in a summer program for gifted high-school students. It can be used in different ways by people in various situations. For anyone studying quantum mechanics, it offers an alternative point of view that I hope will be refreshing. It is designed to give the simplest presentation of the basic topics it covers. I think most people who read *Scientific American* can read this book. It can be read along with broader and more descriptive accounts to learn about the new concepts of the physical world at the basis of quantum mechanics. They are interesting to people outside physics but they are not well known and understood because a complete course in quantum mechanics requires sophisticated mathematics. This book offers an opportunity to actually learn some quantum mechanics, do some problems, and use part of the quantum language, without extensive mathematical preparation. The algebra of complex numbers and matrices is fairly simple and rather fun. The book contains over 100 problems. It also contains references to broader descriptive material.

Presenting quantum mechanics entirely in terms of matrices is not a new idea. Matrix mechanics existed more than half a year before wave mechanics. Born wrote a book on quantum mechanics from the point of view of matrix mechanics alone. It was demolished in a review by Pauli. No single method is sufficient for physicists. They should learn all the different ways of doing quantum mechanics.

Here is one easy way to look rather deeply into quantum mechanics.

THOMAS F. JORDAN

Duluth, Minnesota
June 1985

ACKNOWLEDGMENTS

I want to thank Howard Hanson, Russell Hobbie, Bruce Jones, John Lowenstein, F. A. Matsen, David Mermin, and Roger Stuewer for criticism, suggestions, discussions, encouragement, and help of various kinds.

I am particularly grateful to Professor Eugene Wigner for a conversation about the history of the connection between angular-momentum commutation relations and rotation symmetry.

I thank Charlene Miller for her excellent work in typing the manuscript, making innumerable changes, and obtaining copyright permissions.

Finally, I thank the following publishers for permission to use material from their publications:

1. Charles Scribner's Sons for an excerpt from *Max Born*: *My Life*. (The original English version was first published in 1978 by Taylor & Francis Ltd. and Charles Scribner's Sons.)
2. Walker and Company for material quoted from *The Born–Einstein Letters*, translated by Irene Born. Copyright 1971 by Irene Born.
3. The Open Court Publishing Company for material from *Albert Einstein*: *Philosopher–Scientist* edited by Schilpp.
4. The Nobel Foundation for the English Translation of M. Born's Nobel Lecture.
5. Harper & Row and Allen & Unwin for material from *Physics and Beyond* by W. Heisenberg. Copyright 1971.
6. Academic Press for material from *History of Twentieth Century Physics*, C. Weiner, Ed. Copyright 1977.
7. Gordon and Breach Science Publishers for material from *The Development of Quantum Theory* by P. A. M. Dirac. Copyright 1971.
8. Springer Verlag for material from *The Historical Development of Quantum Theory, Vol. 3, The Formulation of Matrix Mechanics and Its Modifications, 1925–1926* by J. Mehra and H. Rechenberg. Copyright 1982.

T. F. J.

CONTENTS

SUMMARY

Quantum mechanics is a fundamental language of physics. It is used to describe things on a small scale of size where quantities comparable to Planck's constant are important. It represents knowledge gained from many experiments that reveal properties of atoms and atomic particles that are different from what we know in everyday life. Therefore the language is different.

It is a mathematical language. Its character is determined largely by the mathematics that is used. Imaginary and complex numbers are widely used. Physical quantities such as position coordinates, velocities, momenta, angular momenta, and energies are represented by matrices. The algebra of matrices is different from that of numbers because matrices do not generally commute; the product AB of two matrices A and B may be different from BA.

A quantity may have a definite value for a particular state of the physical system or object being described, but for any state there are quantities that do not have definite values. For example, if a position coordinate has a definite value, the momentum in the same direction does not. If a quantity does not have a definite value for a particular state, there are probabilities for finding different possible values, and there is a mean value corresponding to these probabilities. Every quantity has a mean value for each state. The basic rules of quantum mechanics can be stated very simply in terms of these mean values and the matrices that represent physical quantities. They are mostly rules that are natural to follow in using probabilities. Yet we can calculate quite a bit from them without assuming much more.

The simplest example is a spin and magnetic moment described by Pauli matrices. From the multiplication rules for these matrices, and the basic rules of quantum mechanics, we deduce that there are only two possible values μ and $-\mu$ for the projection of the magnetic moment in a given direction. We also deduce that if a projection in one direction has a definite value, a projection in another direction does not; for a projection in a perpendicular direction, there are equal probabilities $\frac{1}{2}$ for the two possible values. A definite value for a projection in one direction is all we can know. It is all that can be measured. From it we can calculate mean values and probabilities for the projections in other directions.

This example is particularly simple because it involves only 2×2 matrices. We study it thoroughly before we consider systems described by bigger matrices. We determine which matrices represent physical quantities related to the spin and magnetic moment, which represent real quantities, and which non-negative real quantities. We investigate their possible values, mean values, and probabilities. We find which real quantities can have definite values together for the same state; they are represented by matrices that commute. All this serves as a model for general rules of quantum mechanics that we use later.

We consider two spins and the corresponding magnetic moments for two particles. We compute mean values of their products for a state where the total spin is zero. From these we compute probabilities for different pairs of values for projections of the two magnetic moments in various directions.

These quantities are measured in experiments. There are differing predictions of what these experiments should find. One is the result of the calculations using quantum mechanics. The other comes from arguments about objective reality and causality that appear to be good common sense. They grew out of Einstein's criticism of quantum mechanics. They lead to predictions called Bell inequalities. We consider one argument that shows quantum mechanics is inconsistent with simple ideas about reality and causality. We also consider an example of a Bell inequality that conflicts with a result of our calculations in quantum mechanics. Experiments agree with quantum mechanics and not with the other predictions. This is a test of quantum mechanics in an area where there was some doubt. It also shows there is something wrong with the common sense about reality and causality that disagrees with quantum mechanics.

The multiplication rules for the Pauli matrices are the key to all our calculations for spins and magnetic moments. The other quantities we consider are position coordinates and momenta of particles and quantities made from them, such as angular momentum and energy. For these the key equations are the commutation relations of the position and momentum matrices.

These commutation relations imply Heisenberg's uncertainty relation. The product of the uncertainties for position and momentum in the same direction cannot be smaller than Planck's constant divided by 4π. We obtain this as a particular case of an uncertainty relation for any matrices that do not commute. We show that the latter follows from the general rules of quantum mechanics.

We consider the energy of an oscillator, a particle that oscillates back and forth along a line. The energy is quantized; it can have only certain discrete values; there is no continuous range of possible values. We see this by looking at a matrix used to represent the energy. We also show that it follows from the formula for the energy in terms of position and momentum and the commutation relation of the position and momentum matrices. This quantization applies to the energy of oscillation of atoms in a molecule. The energy in an atom also is quantized, as described by the Bohr model. That is related to quantization of the orbital angular momentum.

Matrices that represent angular momentum satisfy characteristic commutation relations. Since orbital angular momentum is made from position and momentum, the commutation relations of the position and momentum matrices imply commutation relations for the matrices representing orbital angular momentum. The same commutation relations hold for the spin angular momentum described by Pauli matrices. We show that these commutation relations determine the values angular momentum can have. It is quantized; it can have only certain discrete values; there is no continuous range of possible values. This quantization applies to the energy of rotational motion of atoms in a molecule, which can be expressed in terms of the orbital angular momentum.

We find the possible values for the energy in a hydrogen atom. We use Pauli's method, which reduces the problem to one that is easily solved with the mathematics of angular momentum. We see how different states with the same energy correspond to different values of quantities that can be measured together with the energy.

Quantum mechanics makes a distinction between a physical quantity and its values. Every quantity is represented by a matrix. For each state, some quantities have definite values and others do not. Equations relating different quantities are written in terms of matrices rather than values. This makes a difference.

For example, position and momentum are not quantized. Each has a continuous range of possible values. We can see this from Heisenberg's uncertainty relation. The oscillator energy is just a combination of the squares of the position and momentum. Yet it is quantized. It has no continuous range of possible values. This can happen because the formula for the energy in terms of position and momentum is written in terms of matrices. It could not happen if the formula were written in terms of values.

There are two kinds of equations that relate matrices representing physical quantities. The formulas for energy and orbital angular momentum in terms of position and momentum are examples of one kind. As relations between physical quantities, these equations would be the same without quantum mechanics. The only difference is that in quantum mechanics they are written in terms of matrices.

Examples of the other kind are the commutation relations for position and momentum and for angular momentum. They would make no sense at all without quantum mechanics because they would make no sense if they were written in terms of values rather than matrices. There were no equations like these before quantum mechanics. They are completely new.

To understand the meaning of these new equations, we consider the way physical quantities change when the space and time coordinates change and the way different changes are related. The matrices are used two different ways. They represent the quantities that describe a particular physical system at a given time. They are also used as multipliers to change the matrices that represent physical quantities to describe the system at another time, or at a different location in space, or rotated to a different orientation in space, or moving at a different velocity. This is where the new equations come in.

All these changes correspond to changes of space and time coordinates. They relate descriptions of the same system by observers who use different coordinates. Changes of coordinates can be multiplied. The product of two changes is defined simply as the result of making first one and then the other. Each change of coordinates is represented by a matrix that is used as a multiplier to change the matrices representing physical quantities. A product of matrices that represent changes of coordinates represents the product of the changes of coordinates.

Products of Pauli matrices correspond to products of 180° rotations. The matrices that represent angular momentum are also used to construct the matrices that represent small rotations. The commutation relations that are characteristic of matrices representing angular momentum correspond to multiplication of rotations. The matrices that represent momentum are used to construct the matrices that represent changes of location in space. The commutation relations of position and momentum matrices correspond to the way position coordinates are changed by changes of space location. In the end, when we consider all the different changes of coordinates, we can deduce almost all the commutation relations.

1 A STRANGE EQUATION

Quantum mechanics is the new language physicists use to describe the things the world is made of and how they interact. It is the basic language of atomic, molecular, solid-state, nuclear, and particle physics. Once found, it was developed quickly, then extended and applied with continuing success as each new area of physics grew. It emerged after a quarter century of work in atomic physics in which experiments revealed properties of the atomic world that could not be understood with the existing theories and led physicists into new ways of thinking.

The first steps toward the new language were taken in 1925 by Werner Heisenberg, then Max Born and Pascual Jordan, those three in collaboration, and Paul Dirac. Born was one of the first people to appreciate what was happening. He expected a new mathematical language, a "quantum mechanics," would be needed for atomic physics, and he had the mathematical knowledge to develop it [1–3]. At 42, Born was an established physicist, a professor at Göttingen. Heisenberg, who was 23, had finished his doctoral studies with Arnold Sommerfeld at Munich and had come to Göttingen to work as Born's assistant. Here is part of Born's recollections.[‡]

> In Göttingen we also took part in the attempts to distill the unknown mechanics of the atom out of the experimental results. The logical difficulty became ever more acute. ... The art of guessing correct formulas ... was brought to considerable perfection. ...
>
> This period was brought to a sudden end by Heisenberg .... He ... replaced guesswork by a mathematical rule. ... Heisenberg banished the picture of electron orbits with definite radii and periods of rotation, because these quantities are not observable; he demanded that the theory should be built up by means of quadratic arrays ... of transition probabilities .... To me the

[‡]From Ref. 4. © The Nobel Foundation, 1955.

> decisive part in his work is the requirement that one must find a rule whereby from a given array... the array for the square... may be found (or, in general, the multiplication law of such arrays).
>
> By consideration of ...examples...he found this rule.... This was in the summer of 1925. Heisenberg... took leave of absence... and handed over his paper to me for publication....
>
> Heisenberg's rule of multiplication left me no peace, and after a week of intensive thought and trial, I suddenly remembered an algebraic theory.... Such quadratic arrays are quite familiar to mathematicians and are called matrices, in association with a definite rule of multiplication. I applied this rule to Heisenberg's quantum condition and found that it agreed for the diagonal elements. It was easy to guess what the remaining elements must be, namely, null; and immediately there stood before me the strange formula
>
> $$QP - PQ = \frac{ih}{2\pi}.$$

This is one of the fundamental equations of quantum mechanics. In it Q represents a position coordinate of a particle, P represents the momentum of the particle in the same direction, and i and h are fixed numbers. For an electron in a hydrogen atom, typical values for Q and P are 5×10^{-9} cm and 2×10^{-19} g · cm/s. These are small but otherwise ordinary physical quantities.

Then shouldn't QP be the same as PQ? This equation says they are not the same. That is indeed strange. The number h, which is called Planck's constant, is

$$h \simeq 6.626 \times 10^{-27} \text{ g} \cdot \text{cm}^2/\text{s}.$$

That is very small, so the equation says the difference between QP and PQ is small, but not zero.

There is something else that is strange in this equation. The number i has the property that

$$i^2 = -1$$

so taking the square of both sides of the equation gives

$$(QP - PQ)^2 = \frac{-h^2}{4\pi^2}.$$

Isn't the square of any number positive? How can the square of $QP - PQ$ be negative? We see there are some things that have to be learned before all this can be understood. We will consider the question about squares first and then come back to the question of how QP can be different from PQ.

We can also begin to see that the quantum language gives us a new view of the world. We shall find many features of it that differ from everyday experience and even from common sense. They represent an extension of human knowledge to a much smaller scale of size, to atoms and atomic particles. Nothing as small as Planck's constant would ever be noticed in everyday life. Quantum mechanics is one of the most important and interesting accomplishments of science, but it is not part of our common knowledge. It has been used for over half a century but still, for each of us who learns it, it is strange and wonderful.

REFERENCES

1. M. Born, *Z. Phys. 26*, 379 (1924). An English translation is reprinted in *Sources of Quantum Mechanics*, edited by B. L. van der Waerden. Dover, New York, 1968, pp. 181–198.
2. J. Mehra and H. Rechenberg, *The Historical Development of Quantum Theory*, Volume 2, *The Discovery of Quantum Mechanics 1925*, p. 71, and Volume 3, *The Formulation of Matrix Mechanics and Its Modifications 1925–1926*, p. 44. Springer-Verlag, New York, 1982.
3. D. Serwer, "Unmechanischer Zwang: Pauli, Heisenberg, and the Rejection of the Mechanical Atom, 1923–1925," in *Historical Studies in the Physical Sciences*, Vol. 8, edited by R. McCormmach and L. Pyenson. Johns Hopkins University Press, Baltimore, 1977, particularly pp. 194–195.
4. M. Born, *Science* **122**, 675–679 (1955). This is an English translation of Born's Nobel lecture. I have rewritten the equation in the notation we will use here.

2 IMAGINARY NUMBERS

The number i such that

$$i^2 = -1$$

is called an imaginary number. Inventing it did take some thought and imagination.

Consider the familiar numbers. There are positive numbers such as $\frac{1}{6}$, 1, $\frac{4}{3}$, 2, $\sqrt{2} = 1.4142\ldots$, $\pi = 3.1415\ldots$, the number 0, and negative numbers such as $-\frac{1}{6}$, -1, $-\pi$. Inventing the negative numbers took some imagination too. By -1 we mean the number such that

$$-1 + 1 = 0,$$

$$-1 + 2 = 1,$$

$$-1 + 3 = 2,$$

and so on. The solution of the equation

$$x + 1 = 0$$

is

$$x = -1.$$

If negative numbers were not invented, there would be no solution of this equation.

If the imaginary number i were not invented, there would be no number z that is a solution of the equation

$$z^2 = -1.$$

The positive and negative numbers and zero are all called real numbers. The square of any real number is either zero or positive. It is never negative. For example,

$$(-2)^2 = 4.$$

Therefore i cannot be a real number. It is different from all the real numbers, something new, just as -1 is different from all the positive numbers.

Starting with -1, we can make other negative numbers by multiplying with positive numbers. For example,

$$-1/6 = (1/6)(-1)$$

$$-\pi = \pi(-1).$$

Starting with i, we can make other imaginary numbers by multiplying with real numbers. For example,

$$(1/6)i = (1/6)(i)$$

$$i\sqrt{2} = (\sqrt{2})(i)$$

$$-i = (-1)(i)$$

$$-i\pi = (-\pi)(i).$$

Let y be a real number. Then

$$iy = yi$$

is an imaginary number. Its square is

$$(iy)^2 = i^2y^2 = -y^2,$$

which is negative, or zero if y is zero.

In particular,

$$(-i)^2 = (-1)^2i^2 = -1,$$

so the equation

$$z^2 = -1$$

has two solutions

$$z = i \quad \text{and} \quad z = -i.$$

Let v and y be real numbers. Then iv and iy are imaginary numbers and

$$iv + iy = i(v + y).$$

The sum of two imaginary numbers is an imaginary number. For example,

$$2i + 3i = 5i$$

$$4i - i = 3i.$$

The invention is not complete yet. By multiplying and adding we can make numbers such as

$$(-i)i + i = (-1)(i)(i) + i = 1 + i.$$

Let x and y be real numbers. The number

$$z = x + iy$$

is called a complex number, and x and y are called the real and imaginary parts of z. If x is zero, then z is imaginary. If y is zero, then z is real. Let u and v be real numbers. Then

$$w = u + iv$$

is a complex number, and

$$\begin{aligned} w + z &= u + iv + x + iy \\ &= (u + x) + i(v + y). \end{aligned}$$

The sum of two complex numbers is a complex number. The real part of the sum is the sum of the real parts, and the imaginary part of the sum is the sum of the imaginary parts. For example,

$$(1 - 2i) + (5 + 10i) = 6 + 8i$$

$$(-3 + i) + (2 + 4i) = -1 + 5i$$

The product of the two complex numbers is

$$\begin{aligned} wz &= (u + iv)(x + iy) \\ &= ux + i^2vy + iuy + ivx \\ &= (ux - vy) + i(uy + vx). \end{aligned}$$

For example,

$$(1 - 2i)(5 + 2i) = 5 - 4i^2 + 2i - 10i$$

$$= 9 - 8i$$

$$(-4 + i)(2 + 3i) = -8 + 3i^2 - 12i + 2i$$

$$= -11 - 10i.$$

When we say two complex numbers are the same, we mean that both the real and imaginary parts are the same. Thus

$$w = z$$

or

$$u + iv = x + iy$$

means that

$$u = x$$

and

$$v = y.$$

When we say a complex number is zero, we mean that both the real and imaginary parts are zero. Thus

$$z = 0$$

or

$$x + iy = 0$$

means that

$$x = 0$$

and

$$y = 0.$$

If z is zero, then

$$w + z = (u + 0) + i(v + 0)$$

$$= u + iv = w$$

for any complex number $w = u + iv$.

We write $-z$ for $(-1)z$ and $w - z$ for $w + (-z)$. Thus

$$\begin{aligned} -z &= (-1)x + i(-1)y \\ &= -x + i(-y) \\ &= -x - iy \end{aligned}$$

and

$$z - z = 0$$

so the equation

$$w = z$$

is equivalent to

$$w - z = 0.$$

If $z = x + iy$ is not zero, let

$$z^{-1} = \frac{1}{x^2 + y^2}(x - iy).$$

Then

$$z^{-1}z = \frac{1}{x^2 + y^2}(x - iy)(x + iy) = 1.$$

We write $1/z$ for z^{-1} and w/z for $w(1/z)$. We can add, subtract, multiply, and divide any two complex numbers, except of course we cannot divide by zero. There is nothing surprising in the way these operations are done. All the usual rules of algebra apply. The only new thing is the appearance of imaginary numbers.

For each complex number

$$z = x + iy$$

with x and y real, let

$$z^* = x - iy.$$

This number z^* is called the complex conjugate of z. Evidently

$$(z^*)^* = x + iy = z.$$

From z and z^* we get

$$z + z^* = 2x$$

$$z - z^* = 2iy$$

$$x = \tfrac{1}{2}(z + z^*)$$

$$y = -i\tfrac{1}{2}(z - z^*).$$

If $z = z^*$, then z is real. If $z = -z^*$, then z is imaginary. For example,

$$1^* = 1, \qquad (-1)^* = -1,$$

$$i^* = -i, \qquad (-i)^* = i.$$

Consider again two complex numbers

$$w = u + iv$$

and

$$z = x + iy$$

with u, v, x, y real. For the sum

$$w + z = (u + x) + i(v + y)$$

the complex conjugate is

$$(w + z)^* = (u + x) - i(v + y)$$

$$= (u - iv) + (x - iy)$$

$$= w^* + z^*.$$

The complex conjugate of the sum is the sum of the complex conjugates. For the product

$$wz = (ux - vy) + i(uy + vx)$$

we find that

$$\begin{aligned} w^*z^* &= (u - iv)(x - iy) \\ &= ux - i(-i)vy - iuy - ivx \\ &= (ux - vy) - i(uy + vx) \\ &= (wz)^*. \end{aligned}$$

The complex conjugate of the product is the product of the complex conjugates. When z is not zero we have

$$\begin{aligned} \frac{1}{z} &= \frac{1}{x^2 + y^2}(x - iy) \\ \frac{1}{z^*} &= \frac{1}{x^2 + y^2}(x + iy) \\ &= \left(\frac{1}{z}\right)^*. \end{aligned}$$

Then

$$\left(\frac{w}{z}\right)^* = w^*\left(\frac{1}{z}\right)^* = \frac{w^*}{z^*}.$$

For each complex number

$$z = x + iy$$

with x and y real, let

$$|z| = \sqrt{x^2 + y^2}.$$

This is called the absolute value of z. Evidently

$$|z| \geq 0$$

and $|z|$ is zero only if z is zero. Also

$$\begin{aligned} |z^*| &= |z| \\ z^*z &= (x - iy)(x + iy) \\ &= x^2 - iiy^2 + xiy - iyx \\ &= x^2 + y^2 \\ &= |z|^2. \end{aligned}$$

For the two complex numbers

$$w = u + iv$$

and

$$z = x + iy$$

with u, v, x, y real, we get

$$\begin{aligned} |wz|^2 &= (ux - vy)^2 + (uy + vx)^2 \\ &= u^2x^2 + v^2y^2 + u^2y^2 + v^2x^2 \\ &= (u^2 + v^2)(x^2 + y^2) \\ &= |w|^2|z|^2 \\ |wz| &= |w||z|. \end{aligned}$$

The absolute value of the product is the product of the absolute values. From this we can see that if wz is zero then either w is zero or z is zero, because if $|wz|$ is zero then either $|w|$ is zero or $|z|$ is zero, and $|w|$ is zero only when w is zero.

When z is not zero we have

$$\begin{aligned} \left|\frac{1}{z}\right| &= \frac{1}{x^2 + y^2}\sqrt{x^2 + y^2} \\ &= \frac{\sqrt{x^2 + y^2}}{\sqrt{x^2 + y^2}\sqrt{x^2 + y^2}} \\ &= \frac{1}{\sqrt{x^2 + y^2}} \\ &= \frac{1}{|z|}. \end{aligned}$$

Then

$$\left|\frac{w}{z}\right| = |w|\left|\frac{1}{z}\right| = \frac{|w|}{|z|}.$$

For the sum of two complex numbers w and z we can show that

$$|w + z| \leq |w| + |z|.$$

This follows from

$$|w + z|^2 = (w + z)^*(w + z) = (w^* + z^*)(w + z)$$
$$= w^*w + z^*z + w^*z + z^*w$$
$$= |w|^2 + |z|^2 + 2(\tfrac{1}{2})(w^*z + z^*w)$$

because

$$z^*w = (w^*z)^*$$

so $\frac{1}{2}(w^*z + z^*w)$ is the real part of w^*z and therefore

$$\tfrac{1}{2}(w^*z + z^*w) \leq |w^*z| = |w|\,|z|$$

(just as $x \leq \sqrt{x^2 + y^2}$) and

$$|w + z|^2 \leq |w|^2 + |z|^2 + 2|w|\,|z|$$
$$\leq (|w| + |z|)^2.$$

For each complex number

$$z = x + iy$$

with x, y real, there is a square root

$$\sqrt{z} = \sqrt{\frac{\sqrt{x^2 + y^2}}{2}}\left(\sqrt{1 + \frac{x}{\sqrt{x^2 + y^2}}} + i\sqrt{1 - \frac{x}{\sqrt{x^2 + y^2}}}\right) \quad \text{for} \quad y \geq 0$$

$$\sqrt{z} = \sqrt{\frac{\sqrt{x^2 + y^2}}{2}}\left(-\sqrt{1 + \frac{x}{\sqrt{x^2 + y^2}}} + i\sqrt{1 - \frac{x}{\sqrt{x^2 + y^2}}}\right) \quad \text{for} \quad y \leq 0.$$

We can check that for $y \geq 0$ this gives

$$(\sqrt{z})^2 = \frac{\sqrt{x^2+y^2}}{2}\left(1 + \frac{x}{\sqrt{x^2+y^2}} - \left[1 - \frac{x}{\sqrt{x^2+y^2}}\right]\right.$$

$$\left. + i2\sqrt{1 + \frac{x}{\sqrt{x^2+y^2}}}\,\sqrt{1 - \frac{x}{\sqrt{x^2+y^2}}}\right)$$

$$= x + i\sqrt{x^2+y^2}\sqrt{1 - \frac{x^2}{x^2+y^2}}$$

$$= x + i\sqrt{x^2+y^2}\sqrt{\frac{x^2+y^2-x^2}{x^2+y^2}}$$

$$= x + iy = z.$$

For $y < 0$ it gives the same except for a minus sign with the i, which is

$$(\sqrt{z})^2 = x - i\sqrt{y^2} = x - i(-y) = x + iy = z.$$

For example,

$$\sqrt{1} = 1, \qquad \sqrt{-1} = i.$$

Since

$$(-\sqrt{z})^2 = z,$$

each complex number z has two square roots, $\sqrt{z}$ and $-\sqrt{z}$. It is easy to see there are no others. Suppose

$$w^2 = z.$$

If z is 0, then w is 0. If z is not 0, then $\sqrt{z}$ is not 0; because if $\sqrt{z}$ were 0, its square would be 0. Then

$$\left(\frac{w}{\sqrt{z}}\right)^2 = \frac{w^2}{(\sqrt{z})^2} = \frac{z}{z} = 1,$$

so $w/\sqrt{z}$ is either 1 or -1, which means w is either $\sqrt{z}$ or $-\sqrt{z}$.

Complex numbers are widely used in practical applications as well as in mathematics. There are problems, for example in electrical engineering, for which complex numbers are as useful as negative numbers are for problems with money. In quantum mechanics we shall see that complex numbers are not only useful; they are necessary.

PROBLEMS

Write all answers in the form $x + iy$ with x and y real.

2-1. Work out the following sums and products:
$(3 - i) + (2 + 4i)$,
$(1 + 3i) + 2$,
$(-5 + 2i) - (2 + 2i)$,
$(-2 + i) + (2 + 2i)$,
$(3 - i)(2 + 4i)$,
$(1 + 3i)2$,
$i(1 + 3i)$,
$(-5 + 2i)(2 + 3i)$,
$(2 + 3i)(-2 + 3i)$,
$(2 + 3i)(3 + 2i)$.

2-2. Use the formula for z^{-1} to find $1/i$ and $1/-i$ and check that your answers give $i(1/i) = 1$ and $-i(1/-i) = 1$.

2-3. Find i^*i, $(-i)^*(-i)$, $|i|^2$, $|-i|^2$, $|i|$, $|-i|$.

2-4. Let $z = 3 + 4i$. Find z^*, z^*z, $|z|$, z^2, and $1/z$.

2-5. Show that $|w + z| < |w| + |z|$ for $w = 3$ and $z = 4i$.

2-6. Use the formula for $\sqrt{z}$ to find $\sqrt{i}$ and $\sqrt{-i}$ and then check that your answers give $(\sqrt{i})^2 = i$ and $(\sqrt{-i})^2 = -i$.

2-7. Show that

$$z = -\left(\frac{b}{2a}\right) + \sqrt{\left(\frac{b}{2a}\right)^2 - \frac{c}{a}}$$

and

$$z = -\left(\frac{b}{2a}\right) - \sqrt{\left(\frac{b}{2a}\right)^2 - \frac{c}{a}}$$

are solutions of the quadratic equation

$$az^2 + bz + c = 0$$

for any complex numbers b and c and any complex number a except zero.

3 MATRICES

Now we consider the question of how QP can be different from PQ. The answer, which Born was the first to see, is that they are matrices. A matrix is a square array of complex numbers, for example,

$$\begin{pmatrix} -3 & 5i \\ 0 & 7 \end{pmatrix}.$$

That is a 2×2 matrix. Matrices can be 3×3 or 4×4, or any size. We will work mostly with 2×2 matrices because they are the simplest examples.

When Born realized that Q and P are matrices, he saw not just that they are square arrays of numbers, but that they should be multiplied according to a rule mathematicians had already devised. There are also rules for adding matrices and multiplying them by numbers. These rules of matrix algebra are part of the quantum language.

We use capital letters to denote matrices and small letters for real or complex numbers. The numbers in a matrix are labeled as in

$$A = \begin{pmatrix} a_{11} & a_{12} \\ a_{21} & a_{22} \end{pmatrix}, \qquad B = \begin{pmatrix} b_{11} & b_{12} \\ b_{21} & b_{22} \end{pmatrix}.$$

Thus a_{jk} is the number in the jth row and the kth column of the matrix A.

The sum of the two matrices A and B is the matrix

$$A + B = \begin{pmatrix} a_{11} + b_{11} & a_{12} + b_{12} \\ a_{21} + b_{21} & a_{22} + b_{22} \end{pmatrix}.$$

If

$$C = A + B,$$

then

$$c_{jk} = a_{jk} + b_{jk}.$$

The latter form of the rule applies also to matrices bigger than 2 × 2. For example,

$$\begin{pmatrix} -3 & 5i & -i \\ 0 & 7 & 5 \\ i & 3 & 1 \end{pmatrix} + \begin{pmatrix} 1 & -i & 2 \\ i & 1 & 1 \\ 2 & 1 & -1 \end{pmatrix} = \begin{pmatrix} -2 & 4i & 2-i \\ i & 8 & 6 \\ 2+i & 4 & 0 \end{pmatrix}.$$

Addition of two matrices is defined only if they are the same size. We never add a 2 × 2 matrix to a 3 × 3, for example.

Multiplying the matrix B by the complex number z gives the matrix

$$zB = \begin{pmatrix} zb_{11} & zb_{12} \\ zb_{21} & zb_{22} \end{pmatrix}.$$

If

$$C = zB,$$

then

$$c_{jk} = zb_{jk}.$$

The latter also applies to matrices bigger than 2 × 2. For example,

$$2\begin{pmatrix} -3 & 5i & -i \\ 0 & 7 & 5 \\ i & 3 & 1 \end{pmatrix} = \begin{pmatrix} -6 & 10i & -2i \\ 0 & 14 & 10 \\ 2i & 6 & 2 \end{pmatrix}.$$

In particular,

$$(-1)B = \begin{pmatrix} -b_{11} & -b_{12} \\ -b_{21} & -b_{22} \end{pmatrix}.$$

If

$$C = (-1)B,$$

then

$$c_{jk} = -b_{jk}.$$

We write $-B$ for $(-1)B$ and $A - B$ for $A + (-B)$. Thus subtraction of matrices is defined. For example,

$$\begin{pmatrix} -3 & 5i & -i \\ 0 & 7 & 5 \\ i & 3 & 1 \end{pmatrix} - \begin{pmatrix} 1 & -i & 2 \\ i & 1 & 1 \\ 2 & 1 & -1 \end{pmatrix} = \begin{pmatrix} -4 & 6i & -2-i \\ -i & 6 & 4 \\ -2+i & 2 & 2 \end{pmatrix}.$$

When we say two matrices are equal, we mean they are just the same. If $A = B$, then A and B must be the same size and

$$a_{jk} = b_{jk}$$

for all j, k. For each size, the matrix in which all the numbers are zero is called 0. Thus

$$B = 0$$

means $b_{jk} = 0$ for all j, k. If A and B are the same size and B is zero, then

$$A + B = A.$$

Evidently

$$A - A = 0$$

and $A = B$ is equivalent to

$$A - B = 0.$$

Multiplying any matrix by the number 0 gives the matrix 0,

$$0 \cdot A = 0.$$

All the usual rules of algebra apply to the operations considered so far.

Multiplication is the most interesting part of matrix algebra. The product of the two matrices A and B is the matrix

$$AB = \begin{pmatrix} a_{11}b_{11} + a_{12}b_{21} & a_{11}b_{12} + a_{12}b_{22} \\ a_{21}b_{11} + a_{22}b_{21} & a_{21}b_{12} + a_{22}b_{22} \end{pmatrix}.$$

If

$$C = AB,$$

then

$$c_{jk} = a_{j1}b_{1k} + a_{j2}b_{2k}$$

for 2×2 matrices; this becomes

$$c_{jk} = a_{j1}b_{1k} + a_{j2}b_{2k} + a_{j3}b_{3k}$$

for 3×3 matrices, and

$$c_{jk} = a_{j1}b_{1k} + a_{j2}b_{2k} + \cdots + a_{jn}b_{nk}$$

for $n \times n$ matrices. Multiplication of two matrices is defined only if they are the same size. To get the number in the jth row and the kth column of the product, you combine numbers from the jth row of A and the kth column of B. You can keep track of them with your fingers by letting one finger run across the jth row of A while another runs down the kth column of B. We can write

$$\begin{pmatrix} a_{11} & a_{12} \\ a_{21} & a_{22} \end{pmatrix} \begin{pmatrix} b_{11} & b_{12} \\ b_{21} & b_{22} \end{pmatrix} = \begin{pmatrix} a_{11}b_{11} + a_{12}b_{21} & a_{11}b_{12} + a_{12}b_{22} \\ a_{21}b_{11} + a_{22}b_{21} & a_{21}b_{12} + a_{22}b_{22} \end{pmatrix}.$$

To get the number in the first row and first column of the product, run one finger across the first row of A and another at the same time down the first column of B so they are on a_{11} and b_{11} together and then on a_{12} and b_{21}. Multiplying and adding as you go gives

$$a_{11}b_{11} + a_{12}b_{21}.$$

To get the number in the second row and first column of the product, go across the second row of A and down the first column of B. In general, to get the number in the jth row and kth column of the product, go across the jth row of A and down the kth column of B. This applies also to matrices bigger than 2×2, but the rows and columns are longer. While your fingers keep track of the numbers coming from A and B, you can multiply them together and add to get the number in the product matrix. For example, you can work out these matrix products:

$$\begin{pmatrix} 1 & 2 & 3 \\ 10 & 20 & 30 \\ 100 & 200 & 300 \end{pmatrix}\begin{pmatrix} 1 & 10 & 100 \\ 2 & 20 & 200 \\ 3 & 30 & 300 \end{pmatrix} = \begin{pmatrix} 14 & 140 & 1400 \\ 140 & 1400 & 14{,}000 \\ 1400 & 14{,}000 & 140{,}000 \end{pmatrix}$$

$$\begin{pmatrix} 0 & 1 \\ 1 & 0 \end{pmatrix}\begin{pmatrix} 0 & 1 \\ 1 & 0 \end{pmatrix} = \begin{pmatrix} 1 & 0 \\ 0 & 1 \end{pmatrix}$$

$$\begin{pmatrix} 0 & -i \\ i & 0 \end{pmatrix}\begin{pmatrix} 0 & -i \\ i & 0 \end{pmatrix} = \begin{pmatrix} 1 & 0 \\ 0 & 1 \end{pmatrix}$$

$$\begin{pmatrix} 1 & 0 \\ 0 & -1 \end{pmatrix}\begin{pmatrix} 1 & 0 \\ 0 & -1 \end{pmatrix} = \begin{pmatrix} 1 & 0 \\ 0 & 1 \end{pmatrix}$$

$$\begin{pmatrix} 0 & 1 \\ 1 & 0 \end{pmatrix}\begin{pmatrix} 0 & -i \\ i & 0 \end{pmatrix} = \begin{pmatrix} i & 0 \\ 0 & -i \end{pmatrix}$$

$$\begin{pmatrix} 0 & -i \\ i & 0 \end{pmatrix}\begin{pmatrix} 0 & 1 \\ 1 & 0 \end{pmatrix} = \begin{pmatrix} -i & 0 \\ 0 & i \end{pmatrix}$$

$$\begin{pmatrix} 0 & -i \\ i & 0 \end{pmatrix}\begin{pmatrix} 1 & 0 \\ 0 & -1 \end{pmatrix} = \begin{pmatrix} 0 & i \\ i & 0 \end{pmatrix}$$

$$\begin{pmatrix} 1 & 0 \\ 0 & -1 \end{pmatrix}\begin{pmatrix} 0 & -i \\ i & 0 \end{pmatrix} = \begin{pmatrix} 0 & -i \\ -i & 0 \end{pmatrix}$$

$$\begin{pmatrix} 1 & 0 \\ 0 & -1 \end{pmatrix}\begin{pmatrix} 0 & 1 \\ 1 & 0 \end{pmatrix} = \begin{pmatrix} 0 & 1 \\ -1 & 0 \end{pmatrix}$$

$$\begin{pmatrix} 0 & 1 \\ 1 & 0 \end{pmatrix}\begin{pmatrix} 1 & 0 \\ 0 & -1 \end{pmatrix} = \begin{pmatrix} 0 & -1 \\ 1 & 0 \end{pmatrix}.$$

We see that AB can be different from BA. If AB is the same as BA, we say A and B commute. Some pairs of matrices commute. Others do not. This noncommutative property is what makes matrix multiplication interesting. Except for this, the usual rules of algebra apply to addition and multiplication of matrices and multiplication of matrices by numbers.

Noncommutative algebra was discovered by William Hamilton and Hermann Grassmann independently about 1843. Hamilton had been puzzled for years by a problem essentially the same as the multiplication of 2×2 matrices we just did. A story is told that in response to questions from his young son about what he was doing Hamilton would say he knew how to add but had not yet learned how to multiply, until one day on a walk he suddenly realized the multiplication must be noncommutative and scratched a formula for the multiplication table into the stone of a Dublin bridge.

A few matrices commute with all matrices the same size. One is the zero matrix. Multiplying any matrix B by the matrix 0 gives the matrix 0 either way:

$$0B = 0 = B0;$$

since every number in the matrix 0 is zero, all the numbers in the product matrices $0B$ and $B0$ are zero. We get the same result from multiplying by either the matrix 0 or the number 0, so using the same symbol 0 for both causes no trouble.

The matrix 1 is

$$1 = \begin{pmatrix} 1 & 0 \\ 0 & 1 \end{pmatrix}$$

for 2×2 matrices,

$$1 = \begin{pmatrix} 1 & 0 & 0 \\ 0 & 1 & 0 \\ 0 & 0 & 1 \end{pmatrix}$$

for 3×3 matrices, and

$$1 = \begin{pmatrix} 1 & 0 & \cdots & 0 \\ 0 & 1 & \cdots & 0 \\ \vdots & \vdots & & \vdots \\ 0 & 0 & \cdots & 1 \end{pmatrix}$$

for bigger matrices. For any size, if $A = 1$ then

$$a_{jk} = 1 \qquad \text{for} \quad j = k$$

$$a_{jk} = 0 \qquad \text{for} \quad j \neq k.$$

Then for $C = AB$ we get

$$c_{jk} = a_{jj}b_{jk} = b_{jk}$$

because the other terms in the sum are zero, and for $C = BA$, interchanging the roles of A and B, we get

$$c_{jk} = b_{jk}a_{kk} = b_{jk}.$$

Multiplying any matrix B by the matrix 1 either way gives the matrix B;

$$1B = B = B1.$$

For 2×2 and 3×3 matrices you can check that

$$\begin{pmatrix} 1 & 0 \\ 0 & 1 \end{pmatrix}\begin{pmatrix} b_{11} & b_{12} \\ b_{21} & b_{22} \end{pmatrix} = \begin{pmatrix} b_{11} & b_{12} \\ b_{21} & b_{22} \end{pmatrix} = \begin{pmatrix} b_{11} & b_{12} \\ b_{21} & b_{22} \end{pmatrix}\begin{pmatrix} 1 & 0 \\ 0 & 1 \end{pmatrix}$$

$$\begin{pmatrix} 1 & 0 & 0 \\ 0 & 1 & 0 \\ 0 & 0 & 1 \end{pmatrix}\begin{pmatrix} b_{11} & b_{12} & b_{13} \\ b_{21} & b_{22} & b_{23} \\ b_{31} & b_{32} & b_{33} \end{pmatrix} = \begin{pmatrix} b_{11} & b_{12} & b_{13} \\ b_{21} & b_{22} & b_{23} \\ b_{31} & b_{32} & b_{33} \end{pmatrix}$$

$$= \begin{pmatrix} b_{11} & b_{12} & b_{13} \\ b_{21} & b_{22} & b_{23} \\ b_{31} & b_{32} & b_{33} \end{pmatrix}\begin{pmatrix} 1 & 0 & 0 \\ 0 & 1 & 0 \\ 0 & 0 & 1 \end{pmatrix}.$$

We can add, subtract, and multiply matrices. Can we divide? Let L and M be matrices such that

$$LM = 1 = ML.$$

If K is a matrix such that

$$KM = 1 = MK,$$

then

$$\begin{aligned} K - L &= (K - L)ML \\ &= (KM - LM)L \\ &= (1 - 1)L = 0 \end{aligned}$$

so

$$K = L.$$

The matrix L is unique. We call it M^{-1}, the inverse of M. It is the only matrix M^{-1} such that

$$M^{-1}M = 1 = MM^{-1}.$$

Some matrices have inverses. Others do not. We shall see later which 2×2 matrices have inverses. If M has an inverse, then M^{-1} has an inverse, which is

$$(M^{-1})^{-1} = M.$$

If two matrices M and N have inverses, the product MN also has an inverse, which is

$$(MN)^{-1} = N^{-1}M^{-1},$$

because

$$N^{-1}M^{-1}MN = 1 = MNN^{-1}M^{-1}.$$

If M has an inverse, then AM^{-1} is like A divided by M in that

$$AM^{-1}M = A,$$

but if A does not commute with M then

$$MAM^{-1} \neq A$$

because from

$$MAM^{-1} = A$$

multiplied by M we get

$$MA = AM.$$

PROBLEMS

3-1. Find the matrices for

$$\begin{pmatrix} 1 & 0 \\ 0 & -1 \end{pmatrix} + \begin{pmatrix} 0 & -i \\ i & 0 \end{pmatrix}$$

$$\begin{pmatrix} 0 & 1 \\ 1 & 0 \end{pmatrix} + \begin{pmatrix} 0 & -i \\ i & 0 \end{pmatrix}$$

$$\frac{1}{2}\begin{pmatrix} 1 & 0 \\ 0 & 1 \end{pmatrix} + \frac{1}{2}\begin{pmatrix} 1 & 0 \\ 0 & -1 \end{pmatrix}$$

$$\begin{pmatrix} 0 & 1 \\ 1 & 0 \end{pmatrix} + i\begin{pmatrix} 0 & -i \\ i & 0 \end{pmatrix}.$$

3-2. Compute the matrix products

$$\begin{pmatrix} 1 & -1 & 1 \\ 0 & 1 & 0 \\ 2 & 0 & 3 \end{pmatrix}\begin{pmatrix} 3 & 3 & -1 \\ 0 & 1 & 0 \\ -2 & -2 & 1 \end{pmatrix}$$

$$\begin{pmatrix} 3 & 3 & -1 \\ 0 & 1 & 0 \\ -2 & -2 & 1 \end{pmatrix}\begin{pmatrix} 1 & -1 & 1 \\ 0 & 1 & 0 \\ 2 & 0 & 3 \end{pmatrix}.$$

The answers should be matrices you recognize.

3-3. Find the matrices for these products:

$$\frac{1}{2}\begin{pmatrix} 1 & 1 \\ 1 & 1 \end{pmatrix}\frac{1}{2}\begin{pmatrix} 1 & 1 \\ 1 & 1 \end{pmatrix}$$

$$\frac{1}{2}\begin{pmatrix} 1 & -1 \\ -1 & 1 \end{pmatrix}\frac{1}{2}\begin{pmatrix} 1 & -1 \\ -1 & 1 \end{pmatrix}$$

$$\frac{1}{2}\begin{pmatrix} 1 & 1 \\ 1 & 1 \end{pmatrix}\frac{1}{2}\begin{pmatrix} 1 & -1 \\ -1 & 1 \end{pmatrix}$$

$$\frac{1}{2}\begin{pmatrix} 1 & -1 \\ -1 & 1 \end{pmatrix}\frac{1}{2}\begin{pmatrix} 1 & 1 \\ 1 & 1 \end{pmatrix}.$$

Again the answers should be matrices you recognize. Notice that the product of two matrices can be zero when neither one is zero.

3-4. Show that the inverse of the matrix

$$M = \begin{pmatrix} 1 & 2 & 3 \\ 2 & 4 & 5 \\ 3 & 5 & 6 \end{pmatrix}$$

is

$$M^{-1} = \begin{pmatrix} 1 & -3 & 2 \\ -3 & 3 & -1 \\ 2 & -1 & 0 \end{pmatrix}.$$

3-5. Suppose M has an inverse. Show that if A commutes with M^{-1}, then A commutes with M and, conversely, if A commutes with M, then A commutes with M^{-1}. This means there is an unambiguous matrix $AM^{-1} = M^{-1}A$ for A divided by M if and only if there is an unambiguous matrix $AM = MA$ for the product of A and M.

3-6. Let A, B, and M be matrices the same size and z a complex number. Suppose A and M commute and B and M commute. Show that $A + B$ and M commute, that AB and M commute, and that zB and M commute.

3-7. Show that the inverse of the matrix

$$M = \begin{pmatrix} 2 & 2 - i \\ 2 + i & -2 \end{pmatrix}$$

is $M^{-1} = \frac{1}{9}M$.

4 PAULI MATRICES

The matrices we shall use more than any others are

$$\Sigma_1 = \begin{pmatrix} 0 & 1 \\ 1 & 0 \end{pmatrix},$$

$$\Sigma_2 = \begin{pmatrix} 0 & -i \\ i & 0 \end{pmatrix},$$

$$\Sigma_3 = \begin{pmatrix} 1 & 0 \\ 0 & -1 \end{pmatrix}.$$

They are called the Pauli matrices. We already used them for examples of matrix multiplication. From there we see that

$$\Sigma_1^2 = 1, \qquad \Sigma_2^2 = 1, \qquad \Sigma_3^2 = 1,$$

$$\Sigma_1\Sigma_2 = i\Sigma_3, \qquad \Sigma_2\Sigma_1 = -i\Sigma_3,$$

$$\Sigma_2\Sigma_3 = i\Sigma_1, \qquad \Sigma_3\Sigma_2 = -i\Sigma_1,$$

$$\Sigma_3\Sigma_1 = i\Sigma_2, \qquad \Sigma_1\Sigma_3 = -i\Sigma_2.$$

We use the Pauli matrices and the unit matrix

$$1 = \begin{pmatrix} 1 & 0 \\ 0 & 1 \end{pmatrix}$$

as basic building blocks for 2×2 matrices. For any 2×2 matrix

$$A = \begin{pmatrix} a_{11} & a_{12} \\ a_{21} & a_{22} \end{pmatrix}$$

there are complex numbers z_0, z_1, z_2, z_3 such that

$$A = z_0 1 + z_1 \Sigma_1 + z_2 \Sigma_2 + z_3 \Sigma_3$$

because

$$\begin{pmatrix} a_{11} & a_{12} \\ a_{21} & a_{22} \end{pmatrix} = \begin{pmatrix} z_0 + z_3 & z_1 - iz_2 \\ z_1 + iz_2 & z_0 - z_3 \end{pmatrix}$$

$$= z_0 \begin{pmatrix} 1 & 0 \\ 0 & 1 \end{pmatrix} + z_1 \begin{pmatrix} 0 & 1 \\ 1 & 0 \end{pmatrix} + z_2 \begin{pmatrix} 0 & -i \\ i & 0 \end{pmatrix} + z_3 \begin{pmatrix} 1 & 0 \\ 0 & -1 \end{pmatrix}$$

with

$$a_{11} = z_0 + z_3, \qquad a_{21} = z_1 + iz_2,$$

$$a_{22} = z_0 - z_3, \qquad a_{12} = z_1 - iz_2,$$

$$z_0 = \tfrac{1}{2}(a_{11} + a_{22}), \qquad z_1 = \tfrac{1}{2}(a_{21} + a_{12}),$$

$$z_3 = \tfrac{1}{2}(a_{11} - a_{22}), \qquad z_2 = -i\tfrac{1}{2}(a_{21} - a_{12}).$$

If the matrix A is zero, then all the numbers z_0, z_1, z_2, z_3 must be zero.

To get some experience working with matrices in this form, we now work out some things we will need later. We will use matrices of the form

$$x_1 \Sigma_1 + x_2 \Sigma_2 + x_3 \Sigma_3$$

and

$$y_0 1 + y_1 \Sigma_1 + y_2 \Sigma_2 + y_3 \Sigma_3$$

where x_1, x_2, x_3 and y_0, y_1, y_2, y_3 are real numbers. We find that

$$(x_1 \Sigma_1 + x_2 \Sigma_2 + x_3 \Sigma_3)^2 = (x_1^2 + x_2^2 + x_3^2)1$$

$$(y_0 1 + y_1 \Sigma_1 + y_2 \Sigma_2 + y_3 \Sigma_3)^2 = (y_0^2 + y_1^2 + y_2^2 + y_3^2)1$$

$$+2 y_0 y_1 \Sigma_1 + 2 y_0 y_2 \Sigma_2 + 2 y_0 y_3 \Sigma_3$$

because

$$1^2 = 1, \qquad \Sigma_1^{\,2} = 1,$$

$$\Sigma_2^{\,2} = 1, \qquad \Sigma_3^{\,2} = 1,$$

$$\Sigma_1\Sigma_2 + \Sigma_2\Sigma_1 = 0,$$

$$\Sigma_2\Sigma_3 + \Sigma_3\Sigma_2 = 0,$$

$$\Sigma_3\Sigma_1 + \Sigma_1\Sigma_3 = 0.$$

Consider a 2×2 matrix

$$M = z_0 1 + z_1\Sigma_1 + z_2\Sigma_2 + z_3\Sigma_3.$$

If

$$z_0^{\,2} - z_1^{\,2} - z_2^{\,2} - z_3^{\,2} \neq 0,$$

then M has an inverse, which is

$$M^{-1} = \frac{1}{z_0^{\,2} - z_1^{\,2} - z_2^{\,2} - z_3^{\,2}}\left(z_0 1 - z_1\Sigma_1 - z_2\Sigma_2 - z_3\Sigma_3\right)$$

because this gives

$$M^{-1}M = \frac{1}{z_0^{\,2} - z_1^{\,2} - z_2^{\,2} - z_3^{\,2}}\left(z_0 1 - z_1\Sigma_1 - z_2\Sigma_2 - z_3\Sigma_3\right)$$

$$\times\left(z_0 1 + z_1\Sigma_1 + z_2\Sigma_2 + z_3\Sigma_3\right)$$

$$= \frac{1}{z_0^{\,2} - z_1^{\,2} - z_2^{\,2} - z_3^{\,2}}\left(z_0^{\,2} - z_1^{\,2} - z_2^{\,2} - z_3^{\,2}\right)1 = 1$$

and, similarly,

$$MM^{-1} = 1.$$

Suppose

$$z_0^{\,2} - z_1^{\,2} - z_2^{\,2} - z_3^{\,2} = 0.$$

Then M does not have an inverse. To see this, let

$$N = z_0 1 - z_1 \Sigma_1 - z_2 \Sigma_2 - z_3 \Sigma_3.$$

Then

$$NM = \left(z_0^2 - z_1^2 - z_2^2 - z_3^2\right)1 = 0.$$

If M had an inverse, we would have

$$N = NMM^{-1} = 0M^{-1} = 0,$$

which implies the numbers z_0, z_1, z_2, z_3 are all zero and

$$M = 0.$$

There is no matrix M^{-1} such that

$$M^{-1}0 = 1 = 0M^{-1}.$$

Therefore M does not have an inverse. Thus we have found which 2×2 matrices have inverses, and we have found the inverse for every 2×2 matrix that has one.

If M is a 2×2 matrix that commutes with all three Pauli matrices, then

$$M = z_0 1$$

with z_0 a complex number. To prove this, let

$$M = z_0 1 + z_1 \Sigma_1 + z_2 \Sigma_2 + z_3 \Sigma_3.$$

From

$$M\Sigma_1 = \Sigma_1 M$$

we get

$$z_0 \Sigma_1 + z_1 1 - i z_2 \Sigma_3 + i z_3 \Sigma_2 = z_0 \Sigma_1 + z_1 1 + i z_2 \Sigma_3 - i z_3 \Sigma_2$$

or

$$i2 z_3 \Sigma_2 - i2 z_2 \Sigma_3 = 0,$$

which implies

$$z_2 = 0 = z_3.$$

Similarly,

$$M\Sigma_2 = \Sigma_2 M$$

implies

$$z_3 = 0 = z_1.$$

Which pairs of 2×2 matrices commute? Consider a pair of 2×2 matrices

$$W = w_0 1 + w_1\Sigma_1 + w_2\Sigma_2 + w_3\Sigma_3$$

and

$$Z = z_0 1 + z_1\Sigma_1 + z_2\Sigma_2 + z_3\Sigma_3.$$

It is easy to see that W and Z commute in certain cases. If w_1, w_2, w_3 are all zero, then

$$W = w_0 1. \tag{i}$$

The matrix 1 commutes with any matrix, so in this case W commutes with any 2×2 matrix Z. Similarly, if z_1, z_2, z_3 are all zero, then

$$Z = z_0 1, \tag{ii}$$

which commutes with any 2×2 matrix W. Suppose

$$z_1 = cw_1,$$

$$z_2 = cw_2,$$

$$z_3 = cw_3, \tag{iii}$$

with c a complex number. Then

$$z_1\Sigma_1 + z_2\Sigma_2 + z_3\Sigma_3 = c(w_1\Sigma_1 + w_2\Sigma_2 + w_3\Sigma_3).$$

For any matrix M,

$$M(cM) = (cM)M$$

so the matrices M and cM commute. Therefore the matrices

$$w_1\Sigma_1 + w_2\Sigma_2 + w_3\Sigma_3$$

and

$$z_1\Sigma_1 + z_2\Sigma_2 + z_3\Sigma_3$$

commute. Since the matrices $w_0 1$ and $z_0 1$ commute with any 2×2 matrix, it follows that W and Z commute. These are the only cases where W and Z commute. We have found all the pairs of 2×2 matrices that commute. To prove this, we assume W and Z commute and write out

$$WZ = ZW$$

in terms of the Pauli matrices and 1. This gives

$$(w_1 z_2 - w_2 z_1) i\Sigma_3 + (w_2 z_3 - w_3 z_2) i\Sigma_1 + (w_3 z_1 - w_1 z_3) i\Sigma_2$$

$$= (z_1 w_2 - z_2 w_1) i\Sigma_3 + (z_2 w_3 - z_3 w_2) i\Sigma_1 + (z_3 w_1 - z_1 w_3) i\Sigma_2$$

after the terms that are the same on both sides are canceled. Since each term on the right is minus the corresponding term on the left, this equation is the same as

$$(w_1 z_2 - w_2 z_1) 2i\Sigma_3 + (w_2 z_3 - w_3 z_2) 2i\Sigma_1 + (w_3 z_1 - w_1 z_3) 2i\Sigma_2 = 0,$$

which implies

$$w_1 z_2 - w_2 z_1 = 0,$$

$$w_2 z_3 - w_3 z_2 = 0,$$

$$w_3 z_1 - w_1 z_3 = 0.$$

If w_1, w_2, w_3 are all zero, we have case (i). If, for example, w_1 is not zero, then

$$z_1 = \frac{z_1}{w_1} w_1$$

and from the first and third equations we get

$$z_2 = \frac{z_1}{w_1} w_2,$$

$$z_3 = \frac{z_1}{w_1} w_3,$$

so we have case (iii) with

$$c = \frac{z_1}{w_1}.$$

We get case (iii) similarly if we assume either that w_2 is not zero or that w_3 is not zero. Thus we have proved that every pair of matrices W and Z that commute is included in the cases we listed.

By using the formulas for the squares and products of the Pauli matrices, we have found we can do a variety of calculations algebraically without writing any matrices.

PROBLEMS

4-1. Use the formulas for the squares and products of the Pauli matrices to calculate
$[\frac{1}{2}(1 + \Sigma_1)]^2$, $[\frac{1}{2}(1 - \Sigma_1)]^2$,
$\frac{1}{2}(1 + \Sigma_1)\frac{1}{2}(1 - \Sigma_1)$,
$(i\Sigma_3)\Sigma_1(-i\Sigma_3)$, $(i\Sigma_3)\Sigma_2(-i\Sigma_3)$,
$\sqrt{\frac{1}{2}}(1 + i\Sigma_3)\Sigma_1\sqrt{\frac{1}{2}}(1 - i\Sigma_3)$,
$\sqrt{\frac{1}{2}}(1 + i\Sigma_3)\Sigma_2\sqrt{\frac{1}{2}}(1 - i\Sigma_3)$.
Write your answers in the form $z_0 1 + z_1\Sigma_1 + z_2\Sigma_2 + z_3\Sigma_3$.

4-2. Use the formulas to find the numbers z_0, z_1, z_2, z_3 for the matrix

$$\begin{pmatrix} 0 & 1 \\ 0 & 0 \end{pmatrix}$$

and then check that your answers do give this matrix when you calculate

$$z_0 1 + z_1\Sigma_1 + z_2\Sigma_2 + z_3\Sigma_3.$$

Does this matrix have an inverse?

4-3. Write out the 2×2 matrix for

$$x_1\Sigma_1 + x_2\Sigma_2 + x_3\Sigma_3,$$

multiply the matrix by itself, and thus verify that

$$(x_1\Sigma_1 + x_2\Sigma_2 + x_3\Sigma_3)^2 = (x_1{}^2 + x_2{}^2 + x_3{}^2)1$$

by a second method that is different from the one already used.

4-4. Use the formula to find the inverse M^{-1} for

$$M = \sqrt{\tfrac{1}{2}}\,(1 - i\Sigma_3)$$

and then check that your answer gives

$$M^{-1}M = 1 = MM^{-1}.$$

4-5. Find all the matrices

$$z_0 1 + z_1\Sigma_1 + z_2\Sigma_2 + z_3\Sigma_3$$

that commute with Σ_3.

4-6. Use the formula to find the inverse M^{-1} for

$$M = 2\Sigma_1 + \Sigma_2 + 2\Sigma_3$$

and then check that your answer gives

$$M^{-1}M = 1 = MM^{-1}.$$

Compare this with Problem 3-7.

5 VECTORS

The Pauli matrices are used to represent the spin angular momentum of a particle such as an electron or proton. Angular momentum is a quantity that has both magnitude and direction in three-dimensional space. Such a quantity is called a vector. Velocity is another example of a vector quantity. The velocity of an object has both direction and magnitude, for example east at 50 km per hour.

We can show the direction of a vector by drawing an arrow in three-dimensional space relative to three perpendicular reference directions such as south, east, and up. We can choose any three perpendicular directions for reference axes, but we always label them 1, 2, 3 in an order that allows a map to be laid out in the 1–2 plane so that 1 is south, 2 is east, and 3 is up, not down. We let the length of the arrow represent the magnitude of the vector quantity.

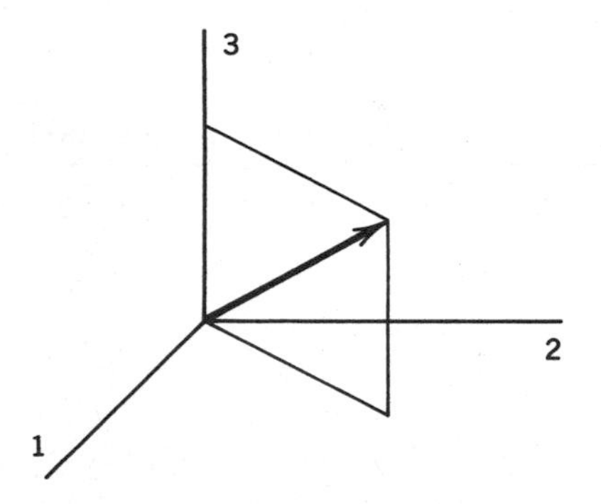

A vector quantity can also be described by its projections in the three perpendicular reference directions. Let x_1, x_2, x_3 be the projections of a vector quantity that has magnitude r.

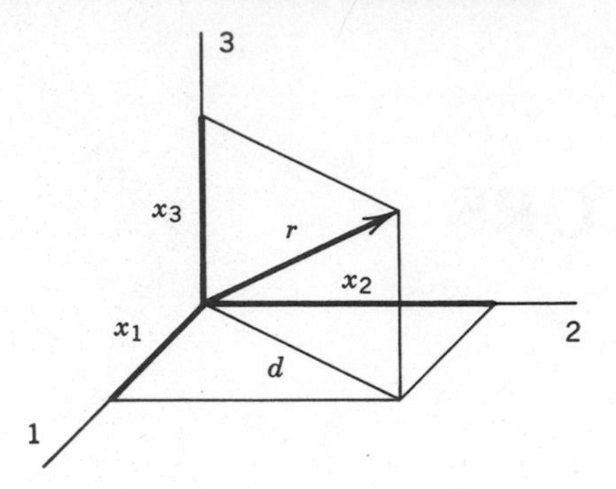

We consider only real quantities now, so the projections x_1, x_2, x_3 are real. The magnitude r is real and non-negative. Then

$$r = \sqrt{x_1^2 + x_2^2 + x_3^2}\,.$$

To see this, look first in the 1–2 plane, where the projection of the vector has length d such that

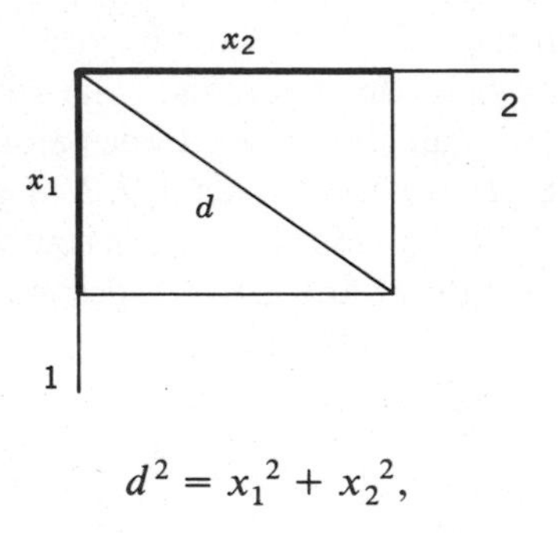

$$d^2 = x_1^2 + x_2^2,$$

then look at the plane through the vector and the 3 axis to see that

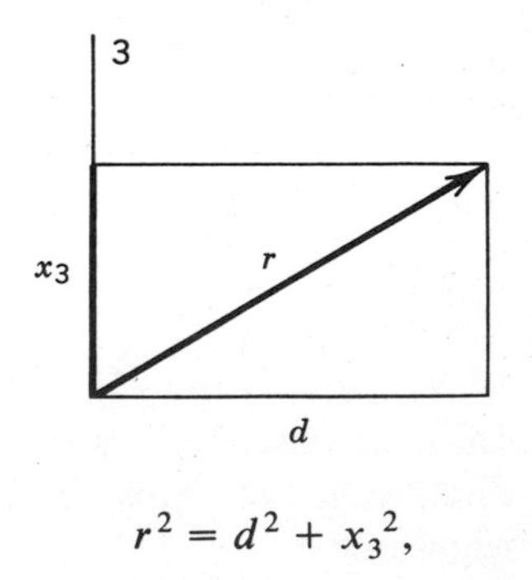

$$r^2 = d^2 + x_3^2,$$

and put the two together to get

$$r^2 = x_1^2 + x_2^2 + x_3^2.$$

Here this is drawn with x_1, x_2, x_3 all positive, but they need not be. A vector can point in any direction, so each projection x_1, x_2, x_3 can be positive, negative, or zero. For example, x_1 could be zero with x_2 negative and x_3 positive.

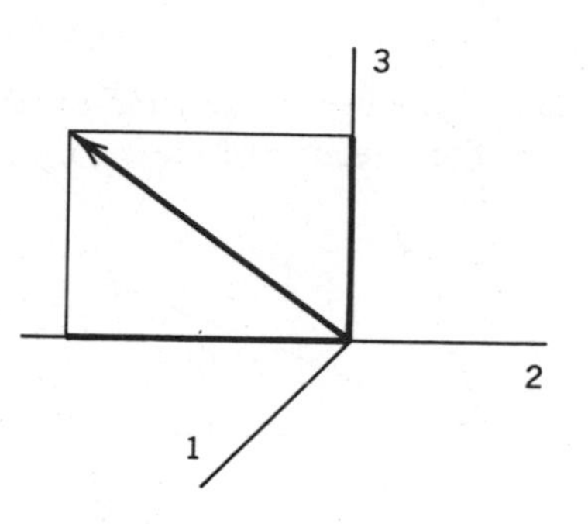

For the example of a velocity vector, the projections are the velocities of motion in the three directions.

For two vectors with projections

$$x_1, x_2, x_3$$

and

$$y_1, y_2, y_3$$

in the three reference directions,

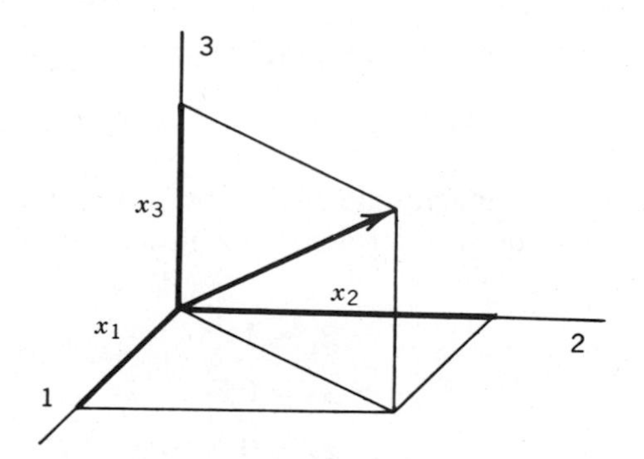

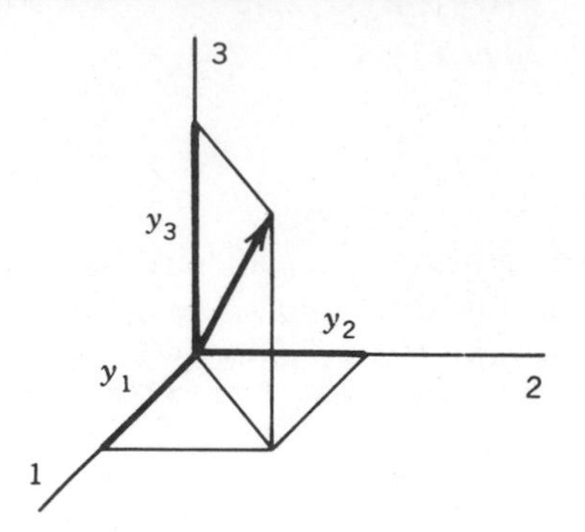

we can look in the plane of the two vectors and consider also the projection of the y_1, y_2, y_3 vector in the direction of the x_1, x_2, x_3 vector.

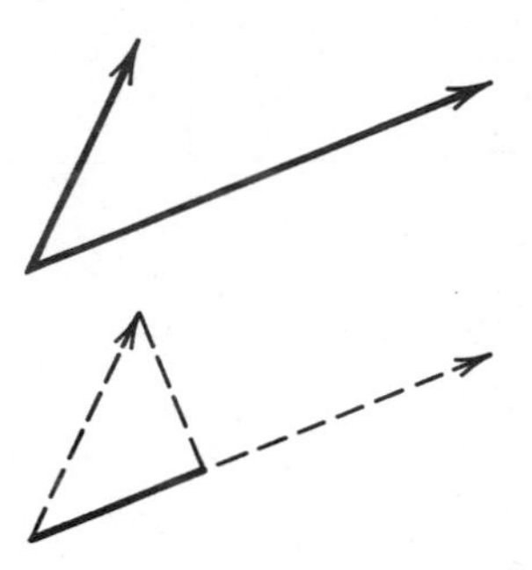

It can be shown that

$$x_1y_1 + x_2y_2 + x_3y_3$$

is the magnitude of the x_1, x_2, x_3 vector times the projection of the y_1, y_2, y_3 vector in the direction of the x_1, x_2, x_3 vector. This is a geometric fact we shall need to use. We shall not prove it here, but we can easily check that it is true in certain cases.

Suppose the two vectors are the same. Then

$$x_1y_1 + x_2y_2 + x_3y_3 = x_1x_1 + x_2x_2 + x_3x_3.$$

This is the square of the magnitude of the vector, which is the magnitude of the vector times the projection of the vector in its own direction.

Suppose

$$x_1 = 1,$$

$$x_2 = 0,$$

$$x_3 = 0.$$

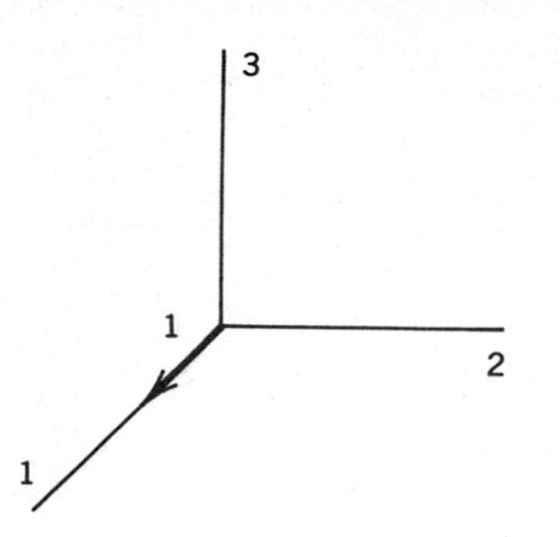

This describes a vector that has direction along the 1 axis and magnitude

$$\sqrt{1^2 + 0^2 + 0^2} = 1.$$

In this case

$$\begin{aligned} x_1 y_1 + x_2 y_2 + x_3 y_3 &= 1 \cdot y_1 + 0 \cdot y_2 + 0 \cdot y_3 \\ &= y_1. \end{aligned}$$

This is the projection of the y_1, y_2, y_3 vector in the 1 direction, which is the magnitude of the x_1, x_2, x_3 vector times the projection of the y_1, y_2, y_3 vector in the direction of the x_1, x_2, x_3 vector. Similarly,

$$x_1 y_1 + x_2 y_2 + x_3 y_3$$

is the projection of the y_1, y_2, y_3 vector in the 2 direction if

$$x_1 = 0,$$

$$x_2 = 1,$$

$$x_3 = 0,$$

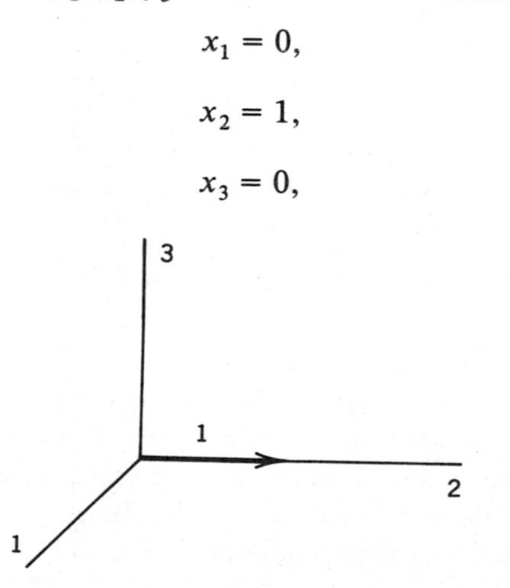

and is the projection of the y_1, y_2, y_3 vector in the 3 direction if

$$x_1 = 0,$$

$$x_2 = 0,$$

$$x_3 = 1.$$

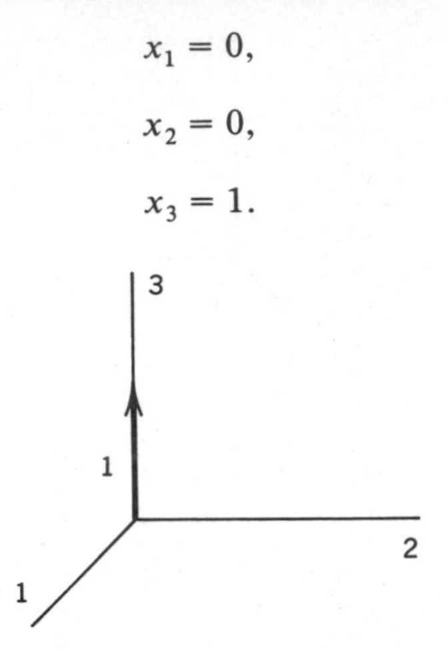

Any direction in three-dimensional space can be specified by real numbers x_1, x_2, x_3 that are the projections of a vector in that direction. Let

$$\sqrt{x_1^2 + x_2^2 + x_3^2} = 1.$$

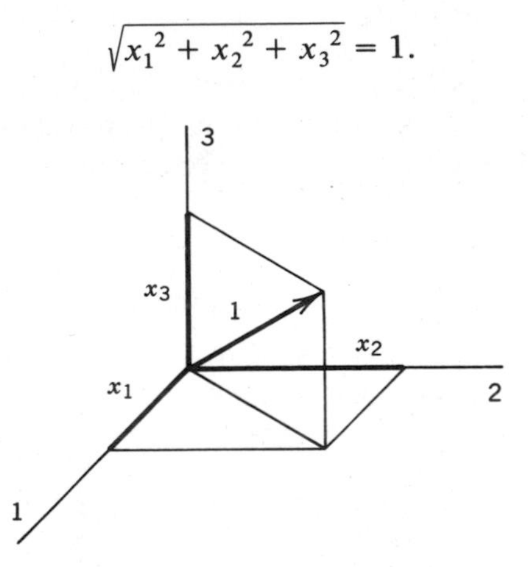

Then the magnitude of the x_1, x_2, x_3 vector is 1 and

$$x_1y_1 + x_2y_2 + x_3y_3$$

is the projection of the y_1, y_2, y_3 vector in the direction of the x_1, x_2, x_3 vector. This is the fact we shall use most often. The first application will be for a vector quantity whose projections in the three reference directions are represented by the Pauli matrices $\Sigma_1, \Sigma_2, \Sigma_3$. Its projection in the direction of the x_1, x_2, x_3 vector is represented by

$$x_1\Sigma_1 + x_2\Sigma_2 + x_3\Sigma_3.$$

PROBLEMS

5-1. Each of the following drawings shows two vectors. The drawing is in the plane of the vectors. Scales show the magnitudes of the vectors. The vectors have projections x_1, x_2, x_3 and y_1, y_2, y_3 in three perpendicular reference directions, but these are not shown. Find

$$x_1y_1 + x_2y_2 + x_3y_3$$

to the nearest whole number in each case:

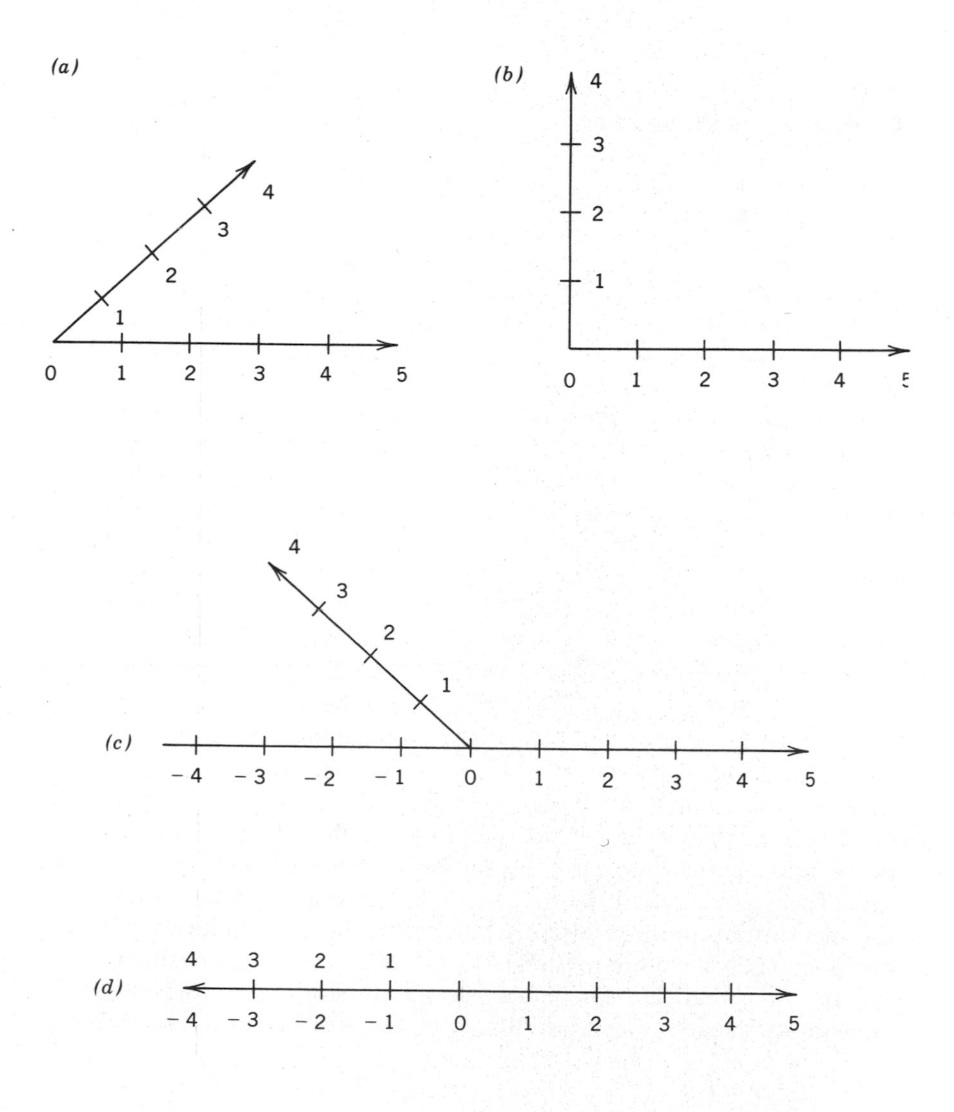

6 PROBABILITY

Consider again the equation

$$QP - PQ = \frac{ih}{2\pi}.$$

Does it seem less strange now? We know something about the imaginary number i. We know that QP can be different from PQ if Q and P are matrices. Still, it is strange that these mathematical inventions describe physical reality.

In this equation, for example, Q represents a position coordinate and P a momentum. If we measure the position and find a value for Q, the value will be a real number. If we measure the momentum and find a value for P, that value will be a real number. These real numbers commute; the value of Q multiplied by the value of P is the same as the value of P multiplied by the value of Q. Furthermore, by multiplying these real numbers and subtracting we cannot get a number whose square is negative. The equation is not consistent with these simple statements about measured values.

In quantum mechanics, the equation is true and the simple ideas about measured values are changed. The position can be measured with unlimited accuracy to get a value for Q, and the momentum can be measured with unlimited accuracy to get a value for P, but both cannot be measured together with unlimited accuracy. Multiplying the two values together has no simple meaning because they refer to two different situations. When the position has been measured, there can be no definite value for the momentum. There are probabilities for various possible momentum values. When the momentum has been measured, there can be no definite value for the position. Then there are probabilities for various possible position values. We shall see that the equation we are considering leads to Heisenberg's uncertainty relation, which puts a limit on the accuracy of measurements of position and momentum together. We shall also work out examples with the Pauli matrices. First we develop the necessary language of probability.

We consider a quantity that can have certain values x. For each possible value x there is a probability $\rho(x)$, which is a number between 0 and 1. It could be 0 or 1. Thus

$$0 \leq \rho(x) \leq 1.$$

The sum of the probabilities for all the possible values x is 1. We write

$$\sum_x \rho(x) = 1.$$

Here Σ_x means sum over all the x. For example, suppose the possible values are

$$x = 6, 12, 24.$$

Then

$$\rho(6) + \rho(12) + \rho(24) = 1.$$

We can imagine various ways this could happen. For example, picture a popcorn machine emitting chaotic bursts of popcorn into a barrel. The bottom of this barrel is divided into three areas that are the mouths of three chutes that take the popcorn to three conveyors that run parallel to one wall of the room at distances 6, 12, and 24. Let x be this distance.

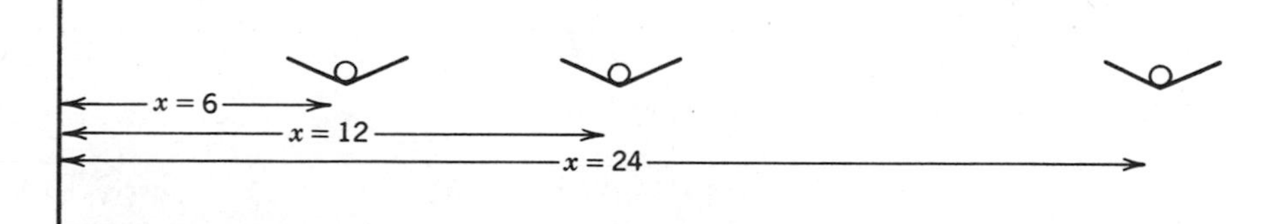

The probability that a piece of popcorn arrives at a particular distance x is proportional to the area of the corresponding chute mouth. If the bottom of the barrel is divided into three equal areas

then

$$\rho(6) = \tfrac{1}{3}, \qquad \rho(12) = \tfrac{1}{3}, \qquad \rho(24) = \tfrac{1}{3}. \tag{i}$$

It may be changed so that, for example:

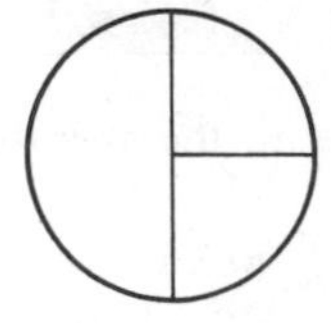

$$\rho(6) = \tfrac{1}{2}, \qquad \rho(12) = \tfrac{1}{4}, \qquad \rho(24) = \tfrac{1}{4}; \tag{ii}$$

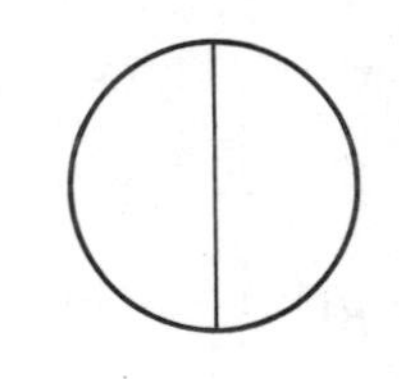

$$\rho(6) = \tfrac{1}{2}, \qquad \rho(12) = \tfrac{1}{2}, \qquad \rho(24) = 0; \tag{iii}$$

or

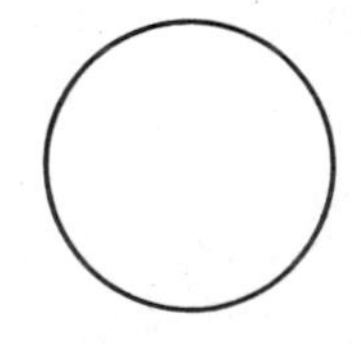

$$\rho(6) = 1, \qquad \rho(12) = 0, \qquad \rho(24) = 0. \tag{iv}$$

If the probability is 1 for one value, then it is 0 for all the others. In this case the quantity has one definite value. It is known with certainty.

The mean of the values x is

$$\langle x \rangle = \sum_x x\rho(x).$$

We refer to this as the mean value of the quantity. It is also called the expectation value, and sometimes simply the average value. In the example it is

$$\langle x \rangle = 6\rho(6) + 12\rho(12) + 24\rho(24).$$

For case (i), with equal probabilities $\frac{1}{3}$, it is

$$\langle x \rangle = 6 \cdot \tfrac{1}{3} + 12 \cdot \tfrac{1}{3} + 24 \cdot \tfrac{1}{3} = 14;$$

for (ii) it is

$$\langle x \rangle = 6 \cdot \tfrac{1}{2} + 12 \cdot \tfrac{1}{4} + 24 \cdot \tfrac{1}{4} = 12;$$

for (iii) it is

$$\langle x \rangle = 6 \cdot \tfrac{1}{2} + 12 \cdot \tfrac{1}{2} + 24 \cdot 0 = 9;$$

and for (iv) it is

$$\langle x \rangle = 6 \cdot 1 + 12 \cdot 0 + 24 \cdot 0 = 6.$$

We use the probabilities $\rho(x)$ to compute the mean values of various quantities we obtain by simple calculation from the original quantity that has values x. If we multiply the original quantity by a fixed number b, we get a quantity that has values bx. It has the value bx if the original quantity has the value x, and the probability for that is $\rho(x)$. Therefore its mean value is

$$\begin{aligned}\langle bx \rangle &= \sum_x bx\rho(x) \\ &= b\sum_x x\rho(x) \\ &= b\langle x \rangle.\end{aligned}$$

In the example half the distance from the wall has values

$$\tfrac{1}{2}x = \tfrac{1}{2}6, \qquad \tfrac{1}{2}12, \qquad \tfrac{1}{2}24$$

and mean value

$$\begin{aligned}\langle \tfrac{1}{2}x \rangle &= \tfrac{1}{2}6\rho(6) + \tfrac{1}{2}12\rho(12) + \tfrac{1}{2}24\rho(24) \\ &= \tfrac{1}{2}[6\rho(6) + 12\rho(12) + 24\rho(24)] \\ &= \tfrac{1}{2}\langle x \rangle.\end{aligned}$$

If we add a fixed number c to the original quantity, we get a quantity that has values $x + c$. It has the value $x + c$ if the original quantity has the value x, and the probability for that is $\rho(x)$. Therefore its mean value is

$$\begin{aligned}\langle x + c\rangle &= \sum_x (x + c)\rho(x)\\ &= \sum_x x\rho(x) + c\sum_x \rho(x)\\ &= \langle x\rangle + c.\end{aligned}$$

In the example, the distance from the first conveyor has values

$$x - 6 = 6 - 6,\quad 12 - 6,\quad 24 - 6$$

and mean value

$$\begin{aligned}\langle x - 6\rangle &= (6 - 6)\rho(6) + (12 - 6)\rho(12) + (24 - 6)\rho(24)\\ &= 6\rho(6) + 12\rho(12) + 24\rho(24) - 6[\rho(6) + \rho(12) + \rho(24)]\\ &= \langle x\rangle - 6.\end{aligned}$$

If we square the original quantity, we get a quantity that has values x^2. It has the value x^2 if the original quantity has the value x, and the probability for the latter is $\rho(x)$. Therefore its mean value is

$$\langle x^2\rangle = \sum_x x^2\rho(x).$$

If we subtract the number $\langle x\rangle$ from the original quantity, we get the quantity that has values $x - \langle x\rangle$. If we square that, we get the quantity that has values $(x - \langle x\rangle)^2$. It has the value $(x - \langle x\rangle)^2$ if the original quantity has the value x, and the probability for the latter is $\rho(x)$. Therefore its mean value is

$$\begin{aligned}\left\langle (x - \langle x\rangle)^2\right\rangle &= \sum_x (x - \langle x\rangle)^2\rho(x)\\ &= \sum_x (x^2 - 2\langle x\rangle x + \langle x\rangle^2)\rho(x)\\ &= \sum_x x^2\rho(x) - 2\langle x\rangle\sum_x x\rho(x) + \langle x\rangle^2\sum_x \rho(x)\\ &= \langle x^2\rangle - 2\langle x\rangle\langle x\rangle + \langle x\rangle^2\\ &= \langle x^2\rangle - \langle x\rangle^2.\end{aligned}$$

Suppose the possible values x of the original quantity are all real numbers. Then $\langle x \rangle$ is a real number, so the numbers $x - \langle x \rangle$ are all real, and the numbers $(x - \langle x \rangle)^2$ are all real and nonnegative. Therefore $\langle (x - \langle x \rangle)^2 \rangle$ is real and

$$\left\langle (x - \langle x \rangle)^2 \right\rangle \geq 0.$$

This implies $\langle x^2 \rangle$ is real and

$$\langle x^2 \rangle \geq \langle x \rangle^2,$$

which implies

$$\langle x^2 \rangle \geq 0.$$

If it is certain that the quantity has a particular value, then $\rho(x)$ is 1 for that value and 0 for all other values. Then $\langle x \rangle$ is that value, and

$$\left\langle (x - \langle x \rangle)^2 \right\rangle = \sum_x (x - \langle x \rangle)^2 \rho(x) = 0$$

because $\rho(x)$ is not 0 only if $x - \langle x \rangle$ is 0. Conversely, if the possible values x are all real and $\langle (x - \langle x \rangle)^2 \rangle$ is 0, then it is certain the quantity has a definite value, which is $\langle x \rangle$, because $(x - \langle x \rangle)^2 \rho(x)$ is nonnegative for all possible values x, so from

$$\sum_x (x - \langle x \rangle)^2 \rho(x) = 0$$

it follows that $\rho(x)$ is 0 unless $x - \langle x \rangle$ is 0. In general, $\langle (x - \langle x \rangle)^2 \rangle$ or $\sqrt{\langle (x - \langle x \rangle)^2 \rangle}$ is a measure of how much the value is likely to differ from $\langle x \rangle$. The square root is called the uncertainty.

In the example,

$$\left\langle (x - \langle x \rangle)^2 \right\rangle = (6 - \langle x \rangle)^2 \rho(6) + (12 - \langle x \rangle)^2 \rho(12) + (24 - \langle x \rangle)^2 \rho(24).$$

For case (i),

$$\left\langle (x - \langle x \rangle)^2 \right\rangle = (6 - 14)^2 \cdot \tfrac{1}{3} + (12 - 14)^2 \cdot \tfrac{1}{3} + (24 - 14)^2 \cdot \tfrac{1}{3} = 56,$$

$$\sqrt{\left\langle (x - \langle x \rangle)^2 \right\rangle} \simeq 7.48.$$

For (ii),

$$\langle (x - \langle x \rangle)^2 \rangle = (6-12)^2 \cdot \tfrac{1}{2} + (12-12)^2 \cdot \tfrac{1}{4} + (24-12)^2 \cdot \tfrac{1}{4} = 54,$$

$$\sqrt{\langle (x - \langle x \rangle)^2 \rangle} \simeq 7.35.$$

For (iii) the value is 6 or 12 with equal probability $\frac{1}{2}$, so $\langle x \rangle$ is 9 and

$$\langle (x - \langle x \rangle)^2 \rangle = (6-9)^2 \cdot \tfrac{1}{2} + (12-9)^2 \cdot \tfrac{1}{2} + (24-9)^2 \cdot 0$$

$$= 9$$

$$\sqrt{\langle (x - \langle x \rangle)^2 \rangle} = 3.$$

For (iv) the value is 6, so $\langle x \rangle$ is 6 and

$$\langle (x - \langle x \rangle)^2 \rangle = (6-6)^2 \cdot 1 + (12-6)^2 \cdot 0 + (24-6)^2 \cdot 0 = 0.$$

We may consider a quantity that has a continuous range of possible values x. Then we divide the continuous range into small intervals, take a representative value x from each interval, and let $\rho(x)$ be the probability that the value is in that interval. For example, if all the numbers between 2 and 3 are possible values, we can consider the representative values

$$x = 2.05, 2.15, 2.25, 2.35, 2.45,$$
$$2.55, 2.65, 2.75, 2.85, 2.95$$

and let $\rho(x)$ be the probability that the value is between $x - .05$ and $x + .05$ so, for example, ρ (2.05) is the probability that the value is between 2 and 2.1.

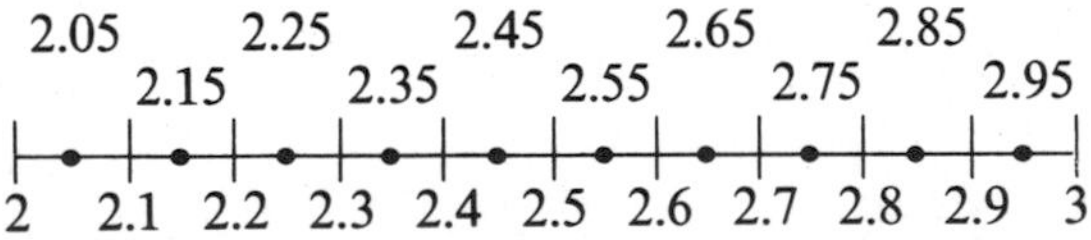

2.05 2.25 2.45 2.65 2.85
2.15 2.35 2.55 2.75 2.95
2 2.1 2.2 2.3 2.4 2.5 2.6 2.7 2.8 2.9 3

All the formulas for mean values can still be used. They are approximately true, and the approximation generally becomes more and more accurate as the intervals are made smaller and smaller.

No measurement can pick one precise value out of a continuous range of possibilities. A measurement is never perfectly accurate. It can only determine that the value is in some small interval. If that interval is larger than the intervals we use for probabilities, the measurement leaves a nonzero probability $\rho(x)$ for more than one representative value x.

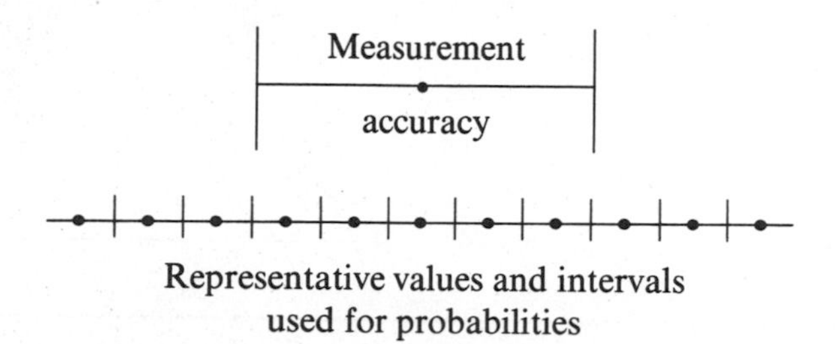

Representative values and intervals used for probabilities

Using smaller and smaller intervals for probabilities insures that this happens. Therefore $\langle (x - \langle x \rangle)^2 \rangle$ is not 0.

We also consider two quantities that can have certain values x and y. For each possible pair of values x, y there is a probability $\mu(x, y)$ such that

$$0 \leq \mu(x, y) \leq 1.$$

The sum of the probabilities for all possible pairs of values x, y is 1. We write

$$\sum_{x, y} \mu(x, y) = 1.$$

For example, suppose the possible pairs of values are

$$(x, y) = (6, 4), (12, 4), (24, 4), (6, 8), (12, 8), (24, 8).$$

Then

$$\mu(6, 4) + \mu(12, 4) + \mu(24, 4) + \mu(6, 8) + \mu(12, 8) + \mu(24, 8) = 1.$$

We can imagine, for example, that each conveyor empties into a barrel that gives the popcorn a probability to go into either of two chutes that empty at distances $y = 4$ and 8 along the line of the conveyor. Then we have a probability for popcorn to arrive at each of six locations described by the coordinates x, y.

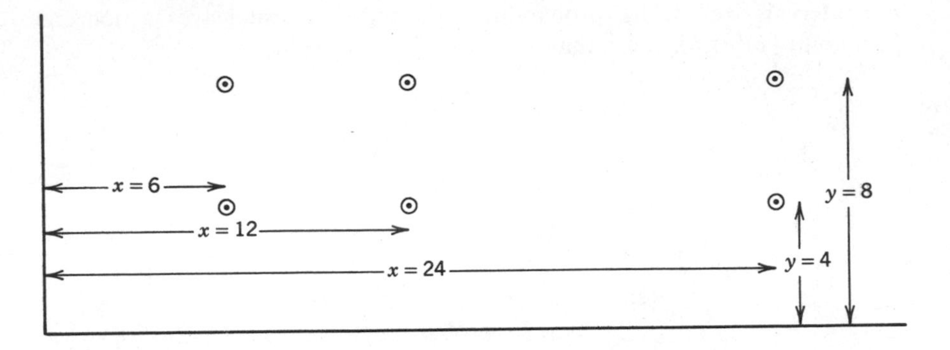

Suppose the probability for popcorn to arrive at each of the conveyors is that of case (ii). Suppose then popcorn at $x = 6$ has probability $\frac{1}{2}$

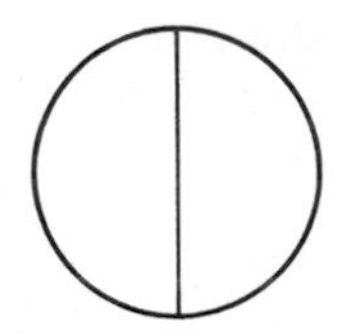

to arrive at either $y = 4$ or $y = 8$, but the popcorn at $x = 12$ all goes to $y = 4$

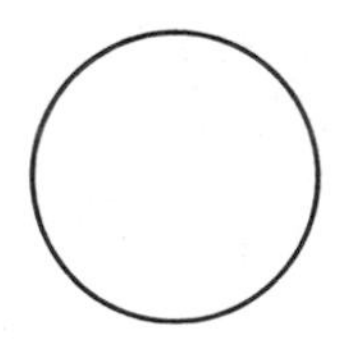

and the popcorn at $x = 24$ all goes to $y = 8$. Then

$$\mu(6,4) = \tfrac{1}{4}, \qquad \mu(12,4) = \tfrac{1}{4}, \qquad \mu(24,4) = 0,$$

$$\mu(6,8) = \tfrac{1}{4}, \qquad \mu(12,8) = 0, \qquad \mu(24,8) = \tfrac{1}{4}.$$

If the two quantities have certain values, then each has a certain value. The first quantity has the value x if the two quantities have the values x, y, and the probability for the latter is $\mu(x, y)$. Therefore the mean value of the first quantity is

$$\langle x \rangle = \sum_{x,y} x\mu(x, y).$$

Similarly, the mean value for the second quantity is

$$\langle y \rangle = \sum_{x,y} y\mu(x, y).$$

In the example, these are

$$\begin{aligned}\langle x \rangle &= 6\mu(6,4) + 12\mu(12,4) + 24\mu(24,4)\\ &\quad + 6\mu(6,8) + 12\mu(12,8) + 24\mu(24,8),\\ \langle y \rangle &= 4\mu(6,4) + 4\mu(12,4) + 4\mu(24,4)\\ &\quad + 8\mu(6,8) + 8\mu(12,8) + 8\mu(24,8).\end{aligned}$$

If we add the two quantities, we get a quantity that has values $x + y$. It has the value $x + y$ if the two quantities have the values x, y, and the probability for the latter is $\mu(x, y)$. Therefore its mean value is

$$\begin{aligned}\langle x + y \rangle &= \sum_{x,y} (x + y)\mu(x, y)\\ &= \sum_{x,y} x\mu(x, y) + \sum_{x,y} y\mu(x, y)\\ &= \langle x \rangle + \langle y \rangle.\end{aligned}$$

We can compute the mean value similarly for any quantity we obtain by simple calculation from the original two quantities that have values x, y. For example, their product has the mean value

$$\langle xy \rangle = \sum_{x,y} xy\mu(x, y).$$

We can work with complex as well as real quantities. If we multiply the second of the original quantities by i and then add the first, we get the complex quantity that has the values

$$z = x + iy.$$

It has the value $x + iy$ if the original quantities have the values x, y, and the probability for that is $\mu(x, y)$. Therefore its mean value is

$$\begin{aligned}\langle z \rangle = \langle x + iy \rangle &= \sum_{x,y} (x + iy)\mu(x, y) \\ &= \sum_{x,y} x\mu(x, y) + i \sum_{x,y} y\mu(x, y) \\ &= \langle x \rangle + i\langle y \rangle.\end{aligned}$$

Suppose the values x, y are all real numbers. Then the square of the absolute value of the complex quantity is the quantity that has values

$$|z|^2 = z^*z = (x - iy)(x + iy) = x^2 + y^2.$$

It has the value $x^2 + y^2$ if the original quantities have the values x, y, and the probability for the latter is $\mu(x, y)$. Therefore its mean value is

$$\begin{aligned}\langle |z|^2 \rangle = \langle z^*z \rangle &= \langle (x - iy)(x + iy) \rangle \\ &= \langle x^2 + y^2 \rangle \\ &= \sum_{x,y} (x^2 + y^2)\mu(x, y) \\ &= \sum_{x,y} x^2\mu(x, y) + \sum_{x,y} y^2\mu(x, y) \\ &= \langle x^2 \rangle + \langle y^2 \rangle.\end{aligned}$$

We see that

$$\langle z^*z \rangle = \langle (x - iy)(x + iy) \rangle \geq 0.$$

In working with complex quantities, we can remember that if a quantity is real, which means its possible values are all real numbers, then its mean value is real. If all the possible values are nonnegative real numbers, then the mean value is real and nonnegative. In particular, $\langle x^2 \rangle$ and $\langle (x - iy)(x + iy) \rangle$ are real and nonnegative if the possible values x, y are all real.

As a particular case of adding a number to a quantity, we get

$$\langle x + 1 \rangle = \langle x \rangle + 1.$$

We can also think of 1 as another quantity that has only the value

$$y = 1,$$

so that

$$\langle x + 1 \rangle = \langle x \rangle + \langle 1 \rangle$$

and

$$\langle 1 \rangle = \sum_{x,y} 1 \cdot \mu(x, y) = \sum_{x,y} \mu(x, y),$$

which implies

$$\langle 1 \rangle = 1.$$

PROBLEMS

6-1. Calculate $\langle x \rangle$, $\langle (x - \langle x \rangle)^2 \rangle$, and $\sqrt{\langle (x - \langle x \rangle)^2 \rangle}$ for these possible values and probabilities:

(a) $x = 4, 8,$
$\rho(4) = \frac{1}{2}$, $\rho(8) = \frac{1}{2}$;

(b) $x = 5, 10,$
$\rho(5) = \frac{4}{5}$, $\rho(10) = \frac{1}{5}$;

(c) $x = -7, -1, 1, 7,$
$\rho(-7) = \frac{1}{4}$, $\rho(-1) = \frac{1}{4}$, $\rho(1) = \frac{1}{4}$, $\rho(7) = \frac{1}{4}$;

(d) $x = -7, -5, -1, 1, 5, 7,$
$\rho(-7) = \frac{1}{6}$, $\rho(-5) = \frac{1}{6}$, $\rho(-1) = \frac{1}{6}$,
$\rho(1) = \frac{1}{6}$, $\rho(5) = \frac{1}{6}$, $\rho(7) = \frac{1}{6}$.

6-2. Use the numbers given in the popcorn example for possible values and probabilities to calculate $\langle x + y \rangle$, $\langle xy \rangle$, and $\langle (x - iy)(x + iy) \rangle$. Check that your answers give

$$\langle x + y \rangle = \langle x \rangle + \langle y \rangle$$

and

$$\langle (x - iy)(x + iy) \rangle \geq 0.$$

Do you get $\langle xy \rangle = \langle x \rangle \langle y \rangle$?

6-3. Calculate $\langle x - \langle x \rangle \rangle$. You can do it without knowing the possible values and probabilities.

6-4. The possible values of a quantity are

$$x = 1, 2, 3.$$

Suppose the probabilities for the first two are

$$\rho(1) = \tfrac{1}{4}, \qquad \rho(2) = \tfrac{1}{2}.$$

Find the probability $\rho(3)$.

6-5. The possible values of a quantity are

$$x = -2, 2.$$

Suppose $\langle x \rangle = 0$. Find the probabilities $\rho(-2)$ and $\rho(2)$.

6-6. The possible values of a quantity are

$$x = -2, 0, 2.$$

Suppose $\langle x \rangle = 0$ and $\langle (x - \langle x \rangle)^2 \rangle = 4$. Find the probabilities $\rho(-2)$, $\rho(0)$, $\rho(2)$.

7 BASIC RULES

A physical object or system of objects is described by quantities such as position, velocity, momentum, angular momentum, and energy. In quantum mechanics, each of these quantities is represented by a matrix. For a particular physical system, all the matrices are the same size, so they can be added and multiplied. Algebraic equations relating different quantities are written in terms of matrices. The equation

$$QP - PQ = \frac{ih}{2\pi}$$

is an example.

Quantum mechanics makes a distinction between a quantity and its values. In any situation, depending on what has been measured, some quantities have definite values and others do not. Therefore algebraic equations relating different quantities are not generally written in terms of their values.

We consider the state of the system in a particular situation, which means all the information about a particular system at a particular time. For each quantity there are probabilities for the different possible values. For a quantity that has a definite value, the probability is 1 for that value and 0 for the other possible values. For a quantity that does not have a definite value, there are nonzero probabilities for more than one possible value. For every quantity there is a mean value corresponding to the probabilities.

Some of the basic rules of quantum mechanics involve simple relations between quantities, expressed in terms of matrices, and corresponding relations between mean values. Consider a quantity represented by a matrix M. Let $\langle M \rangle$ denote its mean value. For any number z, the matrix zM

represents the original quantity multiplied by z. Its mean value is

$$\langle zM \rangle = z\langle M \rangle.$$

The matrix M^2 represents the square of the quantity represented by M. Suppose two quantities are represented by matrices M and N. If the matrix $M + N$ represents a quantity, it has the mean value

$$\langle M + N \rangle = \langle M \rangle + \langle N \rangle.$$

The matrix 1 represents the quantity that has only the value 1. Its mean value is

$$\langle 1 \rangle = 1.$$

For any number w,

$$\begin{aligned} \langle M + w1 \rangle &= \langle M \rangle + \langle w1 \rangle \\ &= \langle M \rangle + w\langle 1 \rangle \\ &= \langle M \rangle + w. \end{aligned}$$

The matrix $M + w1$ represents the original quantity plus w. This allows us to understand the equation

$$(M + w1)^2 = M^2 + 2wM + w^2 1$$

in terms of the original quantity represented by M. In particular, $\sqrt{\langle (M - \langle M \rangle)^2 \rangle}$ is the uncertainty of the quantity for this state. For the matrix $w1$ we sometimes write just w. We often write $M + w$ for $M + w1$, as in $\langle (M - \langle M \rangle)^2 \rangle$.

Some of these rules are more understandable than others. There is no question about zM, for example, but we may question the meaning of $M + N$ when M and N represent quantities that cannot be measured together. These rules involve assumptions. The structure of the quantum language is determined largely by these assumptions, together with the basic assumptions that each quantity is represented by a matrix and has probabilities and a mean value for any state of the system. This structure can be shown to follow from other postulates that have a more direct physical meaning but are expressed in a form that is less often used [1, 2]. Here we state the rules in a form physicists commonly use. That is the way we shall use them. The next step is to see where these rules lead for a system described by 2×2 matrices.

PROBLEMS

7-1. Two quantities are represented by the matrices

$$M = \begin{pmatrix} 3 & 0 & -i \\ 0 & 1 & 0 \\ i & 0 & 3 \end{pmatrix}, \qquad N = \begin{pmatrix} 3 & 0 & 2i \\ 0 & 7 & 0 \\ -2i & 0 & 3 \end{pmatrix}.$$

The possible values of the quantity represented by M are 1, 2, and 4. What are the possible values of the quantity represented by N? Explain how you know that.

REFERENCES

1. J. M. Jauch, *Foundations of Quantum Mechanics*. Addison-Wesley, Reading, Massachusetts, 1968.
2. C. Piron, *Foundations of Quantum Physics*. W. A. Benjamin, Reading, Massachusetts, 1976.

8 SPIN AND MAGNETIC MOMENT

The Pauli matrices Σ_1, Σ_2, Σ_3 are used to represent the spin angular momentum or magnetic moment of an electron or proton or similar particle. These particles are said to have spin $\frac{1}{2}$. We consider only that kind of particle now. The spin angular momentum is a vector quantity. It is analogous to the angular momentum of a spinning top, which is a vector pointing along the axis of rotation. For the atomic particles we are considering, the projections of the spin angular momentum in three perpendicular reference directions are represented by the matrices

$$S_1 = \tfrac{1}{2}\hbar\Sigma_1,$$

$$S_2 = \tfrac{1}{2}\hbar\Sigma_2,$$

$$S_3 = \tfrac{1}{2}\hbar\Sigma_3,$$

where

$$\hbar = \frac{h}{2\pi}$$

$$\simeq 1.0546 \times 10^{-27}\ \text{g} \cdot \text{cm}^2/\text{s}.$$

The spin angular momentum is the angular momentum the particle has even when it is not moving. It is analogous to the angular momentum of a top that is spinning freely in place, with its center of mass stationary. If the particle is moving, it has additional angular momentum. This is called orbital angular momentum. Here we consider only the spin angular momentum and the corresponding magnetic moment.

The magnetic moment of an atomic particle is also a vector quantity. It is analogous to the magnetic moment of a compass needle, which is a vector pointing along the needle. If a spinning top is electrically charged, it has a magnetic moment proportional to its angular momentum. For the atomic particles we are considering, the magnetic moment is proportional to the spin angular momentum. The projections of the magnetic moment in the three perpendicular reference directions are represented by the matrices

$$\mu\Sigma_1, \mu\Sigma_2, \mu\Sigma_3,$$

where μ is a real number. For example,

$$\mu \simeq -9.27 \times 10^{-21} \text{ g} \cdot \text{cm}^2/\text{s}^2 \cdot \text{gauss}$$

for an electron, and

$$\mu \simeq 1.4 \times 10^{-23} \text{ g} \cdot \text{cm}^2/\text{s}^2 \cdot \text{gauss}$$

for a proton. A gauss is a unit of magnetic field strength. The earth's magnetic field is a little less than 1 gauss.

For simplicity we consider the vector quantity whose projections are represented by the matrices $\Sigma_1, \Sigma_2, \Sigma_3$. It is the spin angular momentum divided by $\frac{1}{2}\hbar$. It is also the magnetic moment divided by μ. The magnetic moment is most readily measured.

Magnetic fields exert forces on objects that have magnetic moments. These forces are well understood. They appear to be the same for atomic particles as for macroscopic objects. The rotation of a compass needle in a magnetic field is a familiar example. The magnetic moment rotates to line up with the direction of the magnetic field. If the magnetic field strength is not constant in space, there is also a force that can make the object move as a whole. It accelerates the object along the direction in which the field strength changes. This force is proportional to the projection of the magnetic moment in the direction of the field. It provides a way to measure the projections of the magnetic moment of an atomic particle.

A beam of atomic particles is sent through a magnetic field arranged so its strength varies sharply across the beam. This causes the particles in the beam to be deflected from their original path. These deflections are observed. For each particle the deflection depends on the projection of that particle's magnetic moment in the direction of the field. Thus the projection can be measured. When the direction of the field is changed, the projection in another direction can be measured.

Consider the quantity represented by Σ_1. It is the projection of the magnetic moment in the 1 direction divided by μ. Since

$$\Sigma_1^2 = 1,$$

the square of this quantity can have only the value 1. Therefore the quantity itself can have only the values -1 and 1. Thus the projection of the magnetic moment in the 1 direction can have only the values $-\mu$ and μ.

Similarly, since

$$\Sigma_2^2 = 1$$

and

$$\Sigma_3^2 = 1,$$

each of the quantities represented by Σ_2 and Σ_3 can have only the values -1 and 1. These quantities are the projections in the 2 and 3 directions of the magnetic moment divided by μ. Therefore the projection of the magnetic moment in either the 2 or the 3 direction can have only the values $-\mu$ and μ.

Let x_1, x_2, x_3 be real numbers such that

$$x_1^2 + x_2^2 + x_3^2 = 1.$$

Then

$$\begin{aligned}(x_1\Sigma_1 + x_2\Sigma_2 + x_3\Sigma_3)^2 &= (x_1^2 + x_2^2 + x_3^2)1 \\ &= 1.\end{aligned}$$

Therefore the quantity represented by

$$x_1\Sigma_1 + x_2\Sigma_2 + x_3\Sigma_3$$

can have only the values -1 and 1. This is the projection of the magnetic moment divided by μ in the direction of the x_1, x_2, x_3 vector. Since that could be any direction, it follows that any projection of the magnetic moment, in any direction, can have only the values $-\mu$ and μ.

One might expect that the projection of a vector could have any value in the continuous range between minus the magnitude and the magnitude, depending on the direction of the vector.

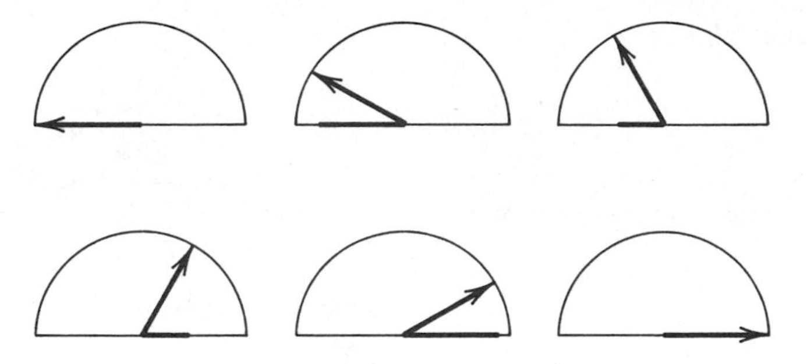

In quantum mechanics this is not so.

There is a clear experimental test. When the magnetic moment is measured, there should be only two values for the deflection acceleration, proportional to the two values $-\mu$ and μ for the projection of the magnetic moment in the direction of the field. A particle can go either way, depending on whether the projection of its magnetic moment is $-\mu$ or μ, but every particle is accelerated by the same amount. Therefore the beam should split in two. The fact that this is observed is strong support for the description of the magnetic moment in terms of 2×2 matrices. The magnitude of μ is determined by precise measurement of the amount of deflection and the velocity and mass of a particle in the beam.

The first experiment of this kind was done by Otto Stern and Walter Gerlach in Frankfurt during 1921 and 1922 [1]. They used a beam of silver atoms and showed clearly that the field split it into two parts corresponding to two values $-\mu$ and μ for the projection of the magnetic moment in the direction of the field, which they measured fairly accurately. We now understand that the magnetic moment they measured for a silver atom is the spin magnetic moment of the single valence electron. At that time it was not understood that the electron has a spin angular momentum and magnetic moment. They had not been untangled from the orbital angular momentum and magnetic moment caused by motion of electrons around the nucleus. It was several years before the development of quantum mechanics, and the Pauli matrices were not yet in use, so the calculations we just did had not been done. However, the Bohr model of the atom had predicted that angular momentum and magnetic moments could have only certain discrete values. That was confirmed in a striking way.

Consider again the quantity represented by Σ_1. It can have only the values -1 and 1. For any state there is a probability $\rho(-1)$ for the value -1 and a probability $\rho(1)$ for the value 1. These are the only possible values, so

$$\rho(-1) + \rho(1) = 1.$$

The mean value is

$$\langle \Sigma_1 \rangle = (-1)\rho(-1) + (1)\rho(1).$$

This is a real number between -1 and 1. Thus

$$-1 \leq \langle \Sigma_1 \rangle \leq 1.$$

By adding and subtracting the two equations, we get

$$\rho(1) = \tfrac{1}{2} + \tfrac{1}{2}\langle \Sigma_1 \rangle \cdot$$

$$\rho(-1) = \tfrac{1}{2} - \tfrac{1}{2}\langle \Sigma_1 \rangle.$$

The probabilities are determined by the mean value. If $\langle \Sigma_1 \rangle$ is 1, then $\rho(1)$ is 1 and $\rho(-1)$ is 0. Then the quantity has the definite value 1. It is certain. If $\langle \Sigma_1 \rangle$ is -1, then $\rho(1)$ is 0 and $\rho(-1)$ is 1. Then the quantity has the definite value -1. That is certain. If $\langle \Sigma_1 \rangle$ is 0, then $\rho(1)$ is $\frac{1}{2}$ and $\rho(-1)$ is $\frac{1}{2}$, so the two possible values are equally probable.

Similarly, the quantities represented by Σ_2 and Σ_3 have mean values $\langle \Sigma_2 \rangle$ and $\langle \Sigma_3 \rangle$ such that

$$-1 \leq \langle \Sigma_2 \rangle \leq 1$$

and

$$-1 \leq \langle \Sigma_3 \rangle \leq 1,$$

and the mean values determine the probabilities for the possible values 1 and -1. All this is the same as for Σ_1 except for the labels 1, 2, 3.

The mean values and probabilities for the quantities represented by $\Sigma_1, \Sigma_2, \Sigma_3$ are not independent. They are limited by the inequality

$$\langle \Sigma_1 \rangle^2 + \langle \Sigma_2 \rangle^2 + \langle \Sigma_3 \rangle^2 \leq 1,$$

which we shall prove now.

Let x_1, x_2, x_3 be three real numbers and let

$$r = \sqrt{x_1^2 + x_2^2 + x_3^2}.$$

Consider the quantity represented by

$$x_1\Sigma_1 + x_2\Sigma_2 + x_3\Sigma_3.$$

Its square is represented by

$$\begin{aligned}(x_1\Sigma_1 + x_2\Sigma_2 + x_3\Sigma_3)^2 &= (x_1^2 + x_2^2 + x_3^2)1 \\ &= (r^2)1\end{aligned}$$

so its square can have only the value r^2. Therefore the quantity itself can have only the values $-r$ and r. Its mean value must be between $-r$ and r. Thus

$$-r \leq \langle x_1\Sigma_1 + x_2\Sigma_2 + x_3\Sigma_3\rangle \leq r.$$

The basic rules imply that

$$\langle x_1\Sigma_1 + x_2\Sigma_2 + x_3\Sigma_3\rangle = x_1\langle\Sigma_1\rangle + x_2\langle\Sigma_2\rangle + x_3\langle\Sigma_3\rangle.$$

Therefore

$$-r \leq x_1\langle\Sigma_1\rangle + x_2\langle\Sigma_2\rangle + x_3\langle\Sigma_3\rangle \leq r.$$

This holds for any real numbers x_1, x_2, x_3. In particular, if

$$x_1 = \langle\Sigma_1\rangle,$$

$$x_2 = \langle\Sigma_2\rangle,$$

$$x_3 = \langle\Sigma_3\rangle,$$

then

$$x_1x_1 + x_2x_2 + x_3x_3 \leq r$$

or

$$r^2 \leq r,$$

which implies

$$r \leq 1$$

and

$$r^2 \leq 1$$

or

$$\langle \Sigma_1 \rangle^2 + \langle \Sigma_2 \rangle^2 + \langle \Sigma_3 \rangle^2 \leq 1.$$

If the projection of the magnetic moment in the 1 direction has the value μ, then

$$\langle \Sigma_1 \rangle = 1.$$

If it has the value $-\mu$, then

$$\langle \Sigma_1 \rangle = -1.$$

In either case

$$\langle \Sigma_1 \rangle^2 = 1.$$

From this and the inequality we just proved, it follows that

$$\langle \Sigma_2 \rangle = 0$$

and

$$\langle \Sigma_3 \rangle = 0.$$

Then for the projection of the magnetic moment in either the 2 direction or the 3 direction, there are equal probabilities $\frac{1}{2}$ for both possible values μ and $-\mu$.

Similarly, if $\langle \Sigma_2 \rangle$ is either 1 or -1, then $\langle \Sigma_3 \rangle$ and $\langle \Sigma_1 \rangle$ are 0; and if $\langle \Sigma_3 \rangle$ is either 1 or -1, then $\langle \Sigma_1 \rangle$ and $\langle \Sigma_2 \rangle$ are 0. If one of the three projections of the magnetic moment has a definite value, either μ or $-\mu$, then for each of the other two there are equal probabilities $\frac{1}{2}$ for both possible values μ and $-\mu$.

In particular, projections of the magnetic moment in two of the perpendicular reference directions cannot both have definite values. If one has a definite value, there are equal probabilities $\frac{1}{2}$ for the two possible values of the other.

These probabilities for more than one possible value cannot be eliminated by measuring more accurately or completely. When a beam of silver atoms is split in two by a magnetic field in the 1 direction, the quantity represented by Σ_1 has the value 1 for every atom in one part of the beam and -1 for every atom in the other part; they correspond to the values μ and $-\mu$ for the projection of the magnetic moment in the 1 direction. For each atom, the quantity represented by Σ_1 is measured with perfect precision when the two possible values 1 and -1 are distinguished by a clean separation of the beam. Probabilities for different possible values are commonly used to describe information from a measurement that is not accurate enough to distinguish those values. There is no need for that here, but there is a new need for the probabilities.

Suppose the quantity represented by Σ_1 has a definite value. Then, according to quantum mechanics, there are probabilities for two possible values for the quantity represented by Σ_2, not because the accuracy is insufficient to distinguish them, but because the quantity represented by Σ_2 does not have a definite value.

Consider successive measurements of the projections of the magnetic moment in the 1 and 2 directions. We refer to these simply as the magnetic moment in the 1 direction and the magnetic moment in the 2 direction. A beam of silver atoms is first split by a magnetic field in the 1 direction. For every atom in one part, the magnetic moment in the 1 direction has the value μ, so

$$\langle \Sigma_1 \rangle = 1,$$

which implies

$$\langle \Sigma_2 \rangle = 0.$$

That part of the beam is immediately split again by a magnetic field in the 2 direction, which separates atoms with different values μ and $-\mu$ of the magnetic moment in the 2 direction. There should be an equal number of each, on the average, because, for an atom entering this field, the fact that $\langle \Sigma_2 \rangle$ is 0 means there are equal probabilities $\frac{1}{2}$ for the two values μ and $-\mu$ of the magnetic moment in the 2 direction.

Now in one part of the beam, every atom has the value μ for the magnetic moment in the 2 direction, so

$$\langle \Sigma_2 \rangle = 1,$$

which implies

$$\langle \Sigma_1 \rangle = 0.$$

This means there are equal probabilities $\frac{1}{2}$ for the two values μ and $-\mu$ of the magnetic moment in the 1 direction. If this part of the beam is

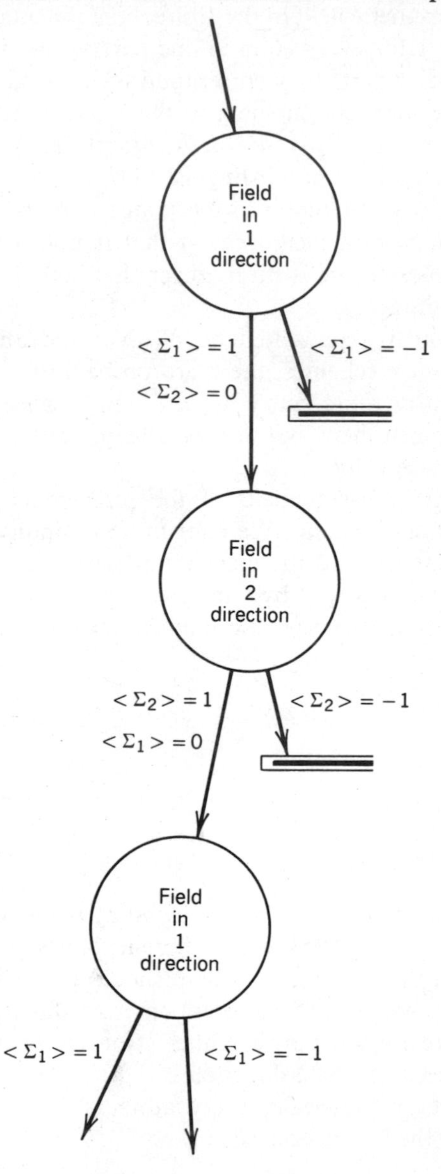

immediately split by another magnetic field in the 1 direction, which separates atoms with different values μ and $-\mu$ of the magnetic moment in the 1 direction, there will be an equal number of each, on the average, even though all these atoms originally came from the beam where the value of the magnetic moment in the 1 direction was known to be μ.

The knowledge obtained by measuring the magnetic moment in the 1 direction is lost when the magnetic moment in the 2 direction is measured. In the end, the probabilities for the magnetic moment in the 1 direction are determined by the fact that $\langle \Sigma_1 \rangle$ must be zero when the magnetic moment in the 2 direction has a definite value. This is not changed by a previous measurement of the magnetic moment in the 1 direction. Here two measurements are no better than one.

PROBLEMS

8-1. Suppose

$$\langle \Sigma_1 \rangle = \tfrac{3}{5}$$

and

$$\langle \Sigma_2 \rangle = \tfrac{4}{5}.$$

Find the six probabilities for the two possible values μ and $-\mu$ for the three projections of the magnetic moment in the perpendicular reference directions.

REFERENCES

1. W. Gerlach and O. Stern, *Z. Phys.* **9**, 349 (1922). An English translation is in *The World of the Atom*, edited by H. A. Boorse and L. Motz. Basic Books, New York, 1966, p. 936.

9 SUNRISE AND FIRST VIEW

When Heisenberg made the first calculations in quantum mechanics, he immediately sensed their significance. He was working alone on the island of Heligoland in the North Sea where he had gone to recover from hay fever. He recalls that at the point where his methods could be tested by calculating a matrix representing energy,

> I became rather excited, and I began to make countless arithmetical errors. As a result, it was almost three o'clock in the morning before the final result of my computations lay before me. The energy principle had held for all the terms, and I could no longer doubt the mathematical consistency and coherence of the kind of quantum mechanics to which my calculations pointed. At first, I was deeply alarmed. I had the feeling that, through the surface of atomic phenomena, I was looking at a strangely beautiful interior and felt almost giddy at the thought that I now had to probe this wealth of mathematical structures nature had so generously spread out before me. I was far too excited to sleep, and so, as a new day dawned, I made for the southern tip of the island, where I had been longing to climb a rock jutting out into the sea. I now did so without too much trouble, and waited for the sun to rise. [1]

Heisenberg's discovery was not the kind that could be announced at a press conference. Few reporters would be interested. Even the people who contributed most quickly and decisively to the development of quantum mechanics needed some time to grasp what Heisenberg had done. His calculations were too complicated to understand at first glance. Born recalls that, "Heisenberg... came to me with a manuscript and asked me to read it.... I remember that I did not read this manuscript at once because I was tired after the term and afraid of hard thinking. But when, after a few days, I did read it, I was fascinated" [2].

Dirac tells this story:

> There were many meetings among the students in Cambridge to discuss scientific problems, and among those there was the Kapitza Club.

> Kapitza... established a club of physicists.... We would meet on Tuesday evenings after dinner.... That was not really a very convenient time for me because I was usually rather sleepy after dinner. I did my work mostly in the morning... and towards the end of the day I was more or less dull, especially after dinner....
>
> In the summer of 1925, Heisenberg came to Cambridge, and he gave a talk to the Kapitza Club. Towards the end... he spoke about some new ideas of his. By that time I was just too exhausted to be able to follow what he said, and I just did not take it in. He was talking about the origins of his ideas of the new mechanics. But I completely failed to realize that he was really introducing something quite revolutionary. Later on I completely forgot what he had said concerning his new theory. I even felt rather convinced that he had not spoken about it at all, but other people who were present at this meeting of the Kapitza Club assured me that he had spoken about it... and I just have to accept that he really did speak about it and that I had failed to respond to it at all, and so missed a great opportunity of getting started on it.
>
> It was a little later when I really got started on the new Heisenberg theory.... Heisenberg sent [his paper] to Fowler.... Fowler sent it on to me with a query, 'What do you think of this?'... At first I was not very much impressed by it. It seemed to me to be too complicated. I just did not see the main point of it, and in particular his derivation of quantum conditions seemed to me to be too far-fetched, so I just put it aside as being of no interest. However, a week or ten days later I returned to this paper of Heisenberg's and studied it more closely. And then I suddenly realized that it did provide the key to the whole solution.... [3]

Our first view of quantum mechanics is simplified immensely when we look at the example of a spin and magnetic moment. Heisenberg's calculations, and all the early work in quantum mechanics, involved infinite matrices. The 2×2 matrices for spin were introduced later. Before we look at infinite matrices and the kind of problems Heisenberg considered, we shall learn as much as we can about quantum mechanics in terms of 2×2 matrices for spins and magnetic moments. This will take us to some interesting recent developments.

REFERENCES

1. W. Heisenberg, *Physics and Beyond*. Harper & Row, New York; George Allen & Unwin (Publishers) Ltd., London, 1971, p. 61.
2. M. Born, *My Life*. Charles Scribner's Sons, New York, 1978, p. 216.
3. P. A. M. Dirac, *History of Twentieth Century Physics*, edited by C. Weiner. Academic Press, New York, 1977, pp. 118–119.

10 ALL QUANTITIES MADE FROM SPIN

Now we make a more systematic and complete study of the quantities represented by 2×2 matrices. We investigate their possible values, mean values, and probabilities. We determine which matrices represent physical quantities related to the spin and magnetic moment. We find which represent real quantities and which represent non-negative real quantities. What we learn here will serve as a model to illustrate general principles of quantum mechanics that we use later for other quantities.

First we shall see that all the real quantities are obtained very simply from the ones we have already considered.

Let x_1, x_2, x_3 be real numbers, let

$$r = \sqrt{x_1^2 + x_2^2 + x_3^2},$$

and let

$$u_1 = \frac{x_1}{r}, \qquad u_2 = \frac{x_2}{r}, \qquad u_3 = \frac{x_3}{r}.$$

Then

$$\begin{aligned} u_1^2 + u_2^2 + u_3^2 &= \frac{x_1^2}{r^2} + \frac{x_2^2}{r^2} + \frac{x_3^2}{r^2} \\ &= \frac{x_1^2 + x_2^2 + x_3^2}{r^2} = \frac{r^2}{r^2} = 1, \end{aligned}$$

and

$$x_1 = ru_1, \qquad x_2 = ru_2, \qquad x_3 = ru_3.$$

Consider the quantity represented by the matrix

$$U = u_1\Sigma_1 + u_2\Sigma_2 + u_3\Sigma_3.$$

It is the projection of the magnetic moment divided by μ in the direction of the u_1, u_2, u_3 vector. Its square is represented by the matrix

$$(u_1\Sigma_1 + u_2\Sigma_2 + u_3\Sigma_3)^2 = (u_1{}^2 + u_2{}^2 + u_3{}^2)1$$
$$= 1,$$

so its square can have only the value 1. Therefore the quantity represented by U can have only the values -1 and 1. Given any state of the system, there are probabilities $\rho(-1)$ for the value -1 and $\rho(1)$ for the value 1. These are the only possible values, so

$$\rho(-1) + \rho(1) = 1.$$

The mean value is

$$\langle U\rangle = (-1)\rho(-1) + (1)\rho(1).$$

By adding and subtracting these equations, we get

$$\rho(1) = \tfrac{1}{2} + \tfrac{1}{2}\langle U\rangle$$
$$\rho(-1) = \tfrac{1}{2} - \tfrac{1}{2}\langle U\rangle.$$

Consider also the quantity represented by the matrix

$$A = x_1\Sigma_1 + x_2\Sigma_2 + x_3\Sigma_3 = ru_1\Sigma_1 + ru_2\Sigma_2 + ru_3\Sigma_3$$
$$= r(u_1\Sigma_1 + u_2\Sigma_2 + u_3\Sigma_3) = rU.$$

It is the first quantity multipled by r. Its mean value is

$$\langle A\rangle = r\langle U\rangle.$$

It can have only the values $-r$ and r, corresponding to the values -1 and 1 for the first quantity. The probability that it has the value $-r$ is the same as the probability $\rho(-1)$ that the first quantity has the value -1, and the probability that it has the value r is the same as the probability $\rho(1)$ that the first quantity has the value 1.

Let x_0 be another real number. Consider the quantity represented by the matrix

$$x_0 1 + A = x_0 1 + x_1\Sigma_1 + x_2\Sigma_2 + x_3\Sigma_3.$$

It is the quantity you get by adding the number x_0 to the quantity represented by the matrix A. Its mean value is

$$\langle x_0 1 + A\rangle = x_0 + \langle A\rangle.$$

Since the quantity represented by A can have only the values $-r$ and r, the quantity represented by $x_0 1 + A$ can have only the values $x_0 - r$ and $x_0 + r$. The probability that it has the value $x_0 - r$ is the same as the probability $\rho(-1)$ that the first quantity has the value -1 and the second the value $-r$, and the probability that it has the value $x_0 + r$ is the same as the probability $\rho(1)$ that the first quantity has the value 1 and the second the value r.

These are real quantities. Their values are real numbers. There are different real quantities for different real numbers x_0, x_1, x_2, x_3. Now we prove there are no other real quantities represented by 2×2 matrices. For any 2×2 matrix M, there are complex numbers z_0, z_1, z_2, z_3 such that

$$M = z_0 1 + z_1\Sigma_1 + z_2\Sigma_2 + z_3\Sigma_3.$$

We show that if M represents a real quantity, all the numbers z_0, z_1, z_2, z_3 are real. If M represents any quantity, real or complex, the basic rules imply its mean value is

$$\langle M\rangle = z_0 + z_1\langle\Sigma_1\rangle + z_2\langle\Sigma_2\rangle + z_3\langle\Sigma_3\rangle.$$

In particular, this is

$$\langle M\rangle = z_0 + z_1$$

for the state where $\langle\Sigma_1\rangle$ is 1 and $\langle\Sigma_2\rangle$ and $\langle\Sigma_3\rangle$ are 0, and it is

$$\langle M\rangle = z_0 - z_1$$

for the state where $\langle\Sigma_1\rangle$ is -1 and $\langle\Sigma_2\rangle$ and $\langle\Sigma_3\rangle$ are 0. If these two mean values are real numbers, then the two numbers

$$z_0 = \tfrac{1}{2}(z_0 + z_1) + \tfrac{1}{2}(z_0 - z_1)$$

and

$$z_1 = \tfrac{1}{2}(z_0 + z_1) - \tfrac{1}{2}(z_0 - z_1)$$

are real. Similarly, we can show that z_2 is real if $\langle M \rangle$ is real for the state where $\langle \Sigma_2 \rangle$ is 1 and $\langle \Sigma_1 \rangle$ and $\langle \Sigma_3 \rangle$ are 0 and the state where $\langle \Sigma_2 \rangle$ is -1 and $\langle \Sigma_1 \rangle$ and $\langle \Sigma_3 \rangle$ are 0, and we can show that z_3 is real if $\langle M \rangle$ is real for the state where $\langle \Sigma_3 \rangle$ is 1 and $\langle \Sigma_1 \rangle$ and $\langle \Sigma_2 \rangle$ are 0 and the state where $\langle \Sigma_3 \rangle$ is -1 and $\langle \Sigma_1 \rangle$ and $\langle \Sigma_2 \rangle$ are 0. The mean value of a real quantity is a real number, because the possible values are real numbers. Therefore, if the matrix M represents a real quantity, all the numbers z_0, z_1, z_2, z_3 are real.

Thus we see that all the real quantities are represented by the matrices $x_0 1 + A$, which we have already investigated.

If K and L are matrices that represent real quantities and b is a real number, then the matrices $K + L$ and bK also represent real quantities. To show this, let

$$K = x_0 1 + x_1 \Sigma_1 + x_2 \Sigma_2 + x_3 \Sigma_3$$

and

$$L = y_0 1 + y_1 \Sigma_1 + y_2 \Sigma_2 + y_3 \Sigma_3.$$

Since the matrices K and L represent real quantities, the numbers x_0, x_1, x_2, x_3 and y_0, y_1, y_2, y_3 are real. Then

$$K + L = (x_0 + y_0)1 + (x_1 + y_1)\Sigma_1 + (x_2 + y_2)\Sigma_2 + (x_3 + y_3)\Sigma_3$$

and

$$bK = (bx_0)1 + (bx_1)\Sigma_1 + (bx_2)\Sigma_2 + (bx_3)\Sigma_3.$$

These represent real quantities, because the number in each ( ) is real.

We have assumed there is a state where $\langle \Sigma_1 \rangle$ is 1 and a state where $\langle \Sigma_1 \rangle$ is -1. We proved that the quantity represented by Σ_1 can have only the values 1 and -1, that these are the only values possible, but we did not prove that there are states where these values occur. We do want to assume this. In general, we assume that for any real numbers x_1, x_2, x_3, there is a state where $\langle U \rangle$ is 1 and a state where $\langle U \rangle$ is -1, because we want to assume that for any direction of the magnetic field in a Stern–Gerlach measurement, there is a state where the magnetic moment in that direction has the value μ and a state where it has the value $-\mu$, so that atoms are deflected in both directions, as observed. These states are basic elements of quantum mechanics for the quantities we are considering. Important properties depend on the assumption that they exist.

We have found all the 2×2 matrices that represent real physical quantities. Do the other 2×2 matrices represent complex quantities? Does every 2×2 matrix represent some real or complex physical quantity?

Consider the matrix

$$\Omega = \tfrac{5}{3}\Sigma_1 + i\tfrac{4}{3}\Sigma_2.$$

Its square is

$$\Omega^2 = \left(\tfrac{25}{9} - \tfrac{16}{9}\right)1 = 1.$$

Therefore a quantity represented by Ω can have only the values -1 and 1. For any state, its mean value must be a real number between -1 and 1. However, the basic rules imply that

$$\langle\Omega\rangle = \tfrac{5}{3}\langle\Sigma_1\rangle + i\tfrac{4}{3}\langle\Sigma_2\rangle.$$

For the state where $\langle\Sigma_2\rangle$ is 1 and $\langle\Sigma_1\rangle$ is 0, this is

$$\langle\Omega\rangle = i\tfrac{4}{3}.$$

The only way to avoid a contradiction is to conclude that this matrix Ω cannot represent any physical quantity at all.

Which matrices represent physical quantities? We show that if a 2×2 matrix M represents a physical quantity there are real numbers x_1, x_2, x_3 and complex numbers z_0 and c such that

$$M = z_0 1 + c(x_1\Sigma_1 + x_2\Sigma_2 + x_3\Sigma_3).$$

We begin, again, with the fact that for any 2×2 matrix M there are complex numbers z_0, z_1, z_2, z_3 such that

$$M = z_0 1 + z_1\Sigma_1 + z_2\Sigma_2 + z_3\Sigma_3.$$

If the matrix M represents a physical quantity, then the matrix

$$\begin{aligned} N &= M - z_0 1 \\ &= z_1\Sigma_1 + z_2\Sigma_2 + z_3\Sigma_3 \end{aligned}$$

also represents a physical quantity. It is the quantity you get by subtracting the complex number z_0 from the quantity represented by M. Let

$$c = \sqrt{z_1^2 + z_2^2 + z_3^2}.$$

Then

$$N^2 = \left(z_1^2 + z_2^2 + z_3^2\right)1 = (c^2)1,$$

so the square of the quantity represented by N can have only the complex value c^2. Therefore the quantity represented by N can have only the complex values $-c$ and c. For each state there is a probability $\rho(-c)$ that the value is $-c$ and a probability $\rho(c)$ that the value is c, and the mean value is

$$\begin{aligned}\langle N\rangle &= (-c)\rho(-c) + c\rho(c)\\ &= c[\rho(c) - \rho(-c)].\end{aligned}$$

On the other hand, the basic rules imply

$$\langle N\rangle = z_1\langle\Sigma_1\rangle + z_2\langle\Sigma_2\rangle + z_3\langle\Sigma_3\rangle.$$

In particular, this is

$$\langle N\rangle = z_1$$

for the state where $\langle\Sigma_1\rangle$ is 1 and $\langle\Sigma_2\rangle$ and $\langle\Sigma_3\rangle$ are 0. Let x_1 be $\rho(c) - \rho(-c)$ for that state. Then x_1 is a real number and

$$z_1 = cx_1.$$

Similarly, we obtain real numbers x_2 and x_3 such that

$$z_2 = cx_2$$

and

$$z_3 = cx_3.$$

Then

$$\begin{aligned}M &= z_0 1 + cx_1\Sigma_1 + cx_2\Sigma_2 + cx_3\Sigma_3\\ &= z_0 1 + c(x_1\Sigma_1 + x_2\Sigma_2 + x_3\Sigma_3).\end{aligned}$$

Thus for every 2×2 matrix M that represents a complex quantity there is a matrix

$$A = x_1\Sigma_1 + x_2\Sigma_2 + x_3\Sigma_3$$

that represents a real quantity such that

$$M = z_0 1 + cA.$$

Each complex quantity is simply related to one of the real quantities we first considered; it is obtained by multiplying the real quantity by a complex number and adding a complex number. The mean value of the complex quantity is

$$\begin{aligned}\langle M\rangle &= z_0 + c\langle A\rangle \\ &= z_0 + cx_1\langle\Sigma_1\rangle + cx_2\langle\Sigma_2\rangle + cx_3\langle\Sigma_3\rangle.\end{aligned}$$

Thus the mean value of any physical quantity for any state can be calculated from the three mean values $\langle\Sigma_1\rangle, \langle\Sigma_2\rangle, \langle\Sigma_3\rangle$ for that state. The complex quantity represented by M has the value $z_0 + cr$ when the real quantity represented by A has the value r, and it has the value $z_0 - cr$ when the real quantity has the value $-r$. The probabilities for these two possibilities are the same as the probabilities for the two values 1 and -1 of the quantity represented by the matrix U corresponding to A, which are

$$\begin{aligned}\rho(1) &= \tfrac{1}{2} + \tfrac{1}{2}\langle u_1\Sigma_1 + u_2\Sigma_2 + u_3\Sigma_3\rangle \\ &= \tfrac{1}{2} + \tfrac{1}{2}(u_1\langle\Sigma_1\rangle + u_2\langle\Sigma_2\rangle + u_3\langle\Sigma_3\rangle)\end{aligned}$$

and

$$\begin{aligned}\rho(-1) &= \tfrac{1}{2} - \tfrac{1}{2}\langle u_1\Sigma_1 + u_2\Sigma_2 + u_3\Sigma_3\rangle \\ &= \tfrac{1}{2} - \tfrac{1}{2}(u_1\langle\Sigma_1\rangle + u_2\langle\Sigma_2\rangle + u_3\langle\Sigma_3\rangle).\end{aligned}$$

Thus all the probabilities for all the physical quantities can be calculated from the three mean values $\langle\Sigma_1\rangle, \langle\Sigma_2\rangle, \langle\Sigma_3\rangle$ for the state.

For any 2×2 matrix M there are real numbers x_0, x_1, x_2, x_3 and y_0, y_1, y_2, y_3 such that

$$\begin{aligned}M &= (x_0 + iy_0)1 + (x_1 + iy_1)\Sigma_1 \\ &\quad + (x_2 + iy_2)\Sigma_2 + (x_3 + iy_3)\Sigma_3 \\ &= x_0 1 + x_1\Sigma_1 + x_2\Sigma_2 + x_3\Sigma_3 \\ &\quad + i(y_0 1 + y_1\Sigma_1 + y_2\Sigma_2 + y_3\Sigma_3).\end{aligned}$$

Thus

$$M = K + iL,$$

where

$$K = x_0 1 + x_1\Sigma_1 + x_2\Sigma_2 + x_3\Sigma_3$$

and

$$L = y_0 1 + y_1\Sigma_1 + y_2\Sigma_2 + y_3\Sigma_3.$$

The matrices K and L represent real physical quantities. Does M represent the complex quantity whose real and imaginary parts are represented by K and L? It does if K and L represent quantities that have definite values together. Then the complex quantity has a definite value when the two real quantities have definite values. Its value is the complex number whose real and imaginary parts are the values of the two real quantities. But if K and L represent quantities that do not have definite values together, there is no complex quantity represented by M. We shall see that K and L represent quantities that do have definite values together if, and only if, M has the form

$$M = z_0 1 + cA,$$

which we have found for a matrix that represents a physical quantity.

For example, the matrix

$$\Omega = \tfrac{5}{3}\Sigma_1 + i\tfrac{4}{3}\Sigma_2$$

cannot represent a physical quantity because the matrices

$$K = \tfrac{5}{3}\Sigma_1$$

and

$$L = \tfrac{4}{3}\Sigma_2$$

represent real quantities that cannot have definite values together.

Each matrix of the form

$$ix_0 1 + i(x_1\Sigma_1 + x_2\Sigma_2 + x_3\Sigma_3)$$

represents an imaginary quantity. Its possible values are $ix_0 + ir$ and $ix_0 - ir$. In particular, the matrix $i\Sigma_2$ represents an imaginary quantity. Its square is represented by the matrix

$$(i\Sigma_2)^2 = -1.$$

Its possible values are i and $-i$. At first glance, it might appear that imaginary numbers could be eliminated by replacing Σ_2 with the matrix

$$\Sigma_2' = i\Sigma_2 = \begin{pmatrix} 0 & 1 \\ -1 & 0 \end{pmatrix}.$$

Then there would be only real numbers in the three matrices $\Sigma_1, \Sigma_2', \Sigma_3$ and we would have

$$\Sigma_1\Sigma_2' = -\Sigma_3, \qquad \Sigma_2'\Sigma_1 = \Sigma_3,$$

$$\Sigma_2'\Sigma_3 = -\Sigma_1, \qquad \Sigma_3\Sigma_2' = \Sigma_1,$$

$$\Sigma_3\Sigma_1 = \Sigma_2', \qquad \Sigma_1\Sigma_3 = -\Sigma_2'.$$

There would be no imaginary numbers in the matrix algebra. However, the values of the quantity represented by Σ_2' are imaginary.

For a matrix that represents a physical quantity, there is a physical meaning for any property of the matrix that can be expressed in terms of the values of the quantity. For example, consider the condition for the matrix to have an inverse. For a matrix

$$M = z_0 1 + cx_1\Sigma_1 + cx_2\Sigma_2 + cx_3\Sigma_3$$

that represents a physical quantity, there is an inverse M^{-1} when

$$z_0^2 - (cx_1)^2 - (cx_2)^2 - (cx_3)^2 \neq 0.$$

This condition can be written as

$$z_0^2 - c^2(x_1^2 + x_2^2 + x_3^2) \neq 0,$$

$$z_0^2 - c^2r^2 \neq 0,$$

or

$$(z_0 + cr)(z_0 - cr) \neq 0,$$

which means neither $z_0 + cr$ nor $z_0 - cr$ is 0. Thus a matrix that represents a physical quantity has an inverse when neither of the possible values of the quantity is 0.

PROBLEMS

10-1. Find the possible values for the quantity represented by the matrix

$$\tfrac{1}{2}(1 + \Sigma_1).$$

Do this two different ways. First, consider this matrix as a particular case of

$$x_0 1 + x_1 \Sigma_1 + x_2 \Sigma_2 + x_3 \Sigma_3$$

for which we know the possible values. Second, consider what the equation

$$\left[\tfrac{1}{2}(1 + \Sigma_1)\right]^2 = \tfrac{1}{2}(1 + \Sigma_1)$$

tells you about the possible values. Check that you get the same answer both ways.

10-2. For each of the matrices (a)–(f):

(i) determine whether the matrix represents a physical quantity; if it does, then
(ii) find the possible values of the quantity;
(iii) determine whether the quantity is real; and
(iv) determine whether the matrix has an inverse.

(a) $2 + 2\Sigma_1$
(b) $3\Sigma_1 + 4\Sigma_2$
(c) $5 + 3\Sigma_1 + 4\Sigma_2$
(d) $i + 3\Sigma_1 + 4\Sigma_2$
(e) $i(3\Sigma_1 + 4\Sigma_2)$
(f) $5\Sigma_1 + i4\Sigma_2$

10-3. Suppose K and L are 2×2 matrices that represent real physical quantities. Does KL represent a real physical quantity? Either show that it does or find an example where it does not.

10-4. Find the possible values for the quantity represented by the matrix

$$\frac{1}{2}\begin{pmatrix} 1 & -1 \\ -1 & 1 \end{pmatrix}.$$

10-5. Find the possible values of the quantity represented by the matrix

$$2\Sigma_1 + \Sigma_2 + 2\Sigma_3.$$

Compare this with Problem 4-6.

11 NON-NEGATIVE QUANTITIES

The matrices that represent non-negative real quantities will play an important role in our development of the general rules for quantum mechanics. Here we establish these rules for the 2×2 matrices that represent the non-negative real quantities related to the spin and magnetic moment.

Consider a real quantity represented by a matrix

$$K = x_0 1 + x_1 \Sigma_1 + x_2 \Sigma_2 + x_3 \Sigma_3.$$

Its possible values are $x_0 + r$ and $x_0 - r$, where

$$r = \sqrt{x_1^2 + x_2^2 + x_3^2}.$$

A real quantity is called non-negative if all its possible values are non-negative. For the quantity represented by K, this means

$$x_0 + r \geq 0 \qquad \text{and} \qquad x_0 - r \geq 0$$

or

$$x_0 \geq r.$$

The square of any real quantity is non-negative. Consider a real quantity represented by a matrix

$$L = y_0 1 + y_1 \Sigma_1 + y_2 \Sigma_2 + y_3 \Sigma_3.$$

Its square is represented by the matrix

$$L^2 = (y_0^2 + y_1^2 + y_2^2 + y_3^2) 1 + 2 y_0 y_1 \Sigma_1 + 2 y_0 y_2 \Sigma_2 + 2 y_0 y_3 \Sigma_3.$$

This is a non-negative quantity. To prove it, let

$$K = L^2.$$

That means

$$x_0 = y_0^2 + y_1^2 + y_2^2 + y_3^2,$$

$$x_1 = 2y_0y_1,$$

$$x_2 = 2y_0y_2,$$

$$x_3 = 2y_0y_3.$$

Then

$$x_0 \geq 0$$

and

$$x_0^2 \geq r^2$$

because

$$(y_0^2 + y_1^2 + y_2^2 + y_3^2)^2 \geq (2y_0y_1)^2 + (2y_0y_2)^2 + (2y_0y_3)^2$$

because

$$(y_0^2)^2 + (y_1^2 + y_2^2 + y_3^2)^2 - 2y_0^2y_1^2 - 2y_0^2y_2^2 - 2y_0^2y_3^2 \geq 0$$

because

$$(y_0^2 - y_1^2 - y_2^2 - y_3^2)^2 \geq 0.$$

Therefore

$$x_0 \geq r,$$

which means K represents a non-negative quantity. Thus for any matrix L that represents a real quantity, the matrix L^2 represents a non-negative real quantity.

Conversely, every non-negative real quantity is the square of a real quantity. Suppose the matrix K represents a non-negative quantity. Then

$$x_0 \geq r.$$

Let

$$L = \sqrt{\tfrac{1}{2}}\left[\left(\sqrt{x_0 + \sqrt{x_0^2 - r^2}}\right)1 + \frac{1}{\sqrt{x_0 + \sqrt{x_0^2 - r^2}}}(x_1\Sigma_1 + x_2\Sigma_2 + x_3\Sigma_3)\right].$$

Then L represents a real quantity, and

$$L^2 = \frac{1}{2}\left(x_0 + \sqrt{x_0^2 - r^2} + \frac{x_1^2 + x_2^2 + x_3^2}{x_0 + \sqrt{x_0^2 - r^2}}\right)1 + x_1\Sigma_1 + x_2\Sigma_2 + x_3\Sigma_3$$

$$= \frac{1}{2}\left(\frac{x_0^2 + 2x_0\sqrt{x_0^2 - r^2} + x_0^2 - r^2 + x_1^2 + x_2^2 + x_3^2}{x_0 + \sqrt{x_0^2 - r^2}}\right)1$$

$$+ x_1\Sigma_1 + x_2\Sigma_2 + x_3\Sigma_3$$

$$= \frac{1}{2}\left(\frac{2x_0^2 + 2x_0\sqrt{x_0^2 - r^2}}{x_0 + \sqrt{x_0^2 - r^2}}\right)1 + x_1\Sigma_1 + x_2\Sigma_2 + x_3\Sigma_3$$

$$= x_0 1 + x_1\Sigma_1 + x_2\Sigma_2 + x_3\Sigma_3$$

$$= K.$$

Thus for any matrix K that represents a non-negative quantity, there is a matrix L that represents a real quantity such that

$$K = L^2.$$

If B and D are matrices that represent real quantities, then the matrix

$$(B + iD)(B - iD)$$

represents a non-negative real quantity. To prove this let

$$B = b_0 1 + b_1\Sigma_1 + b_2\Sigma_2 + b_3\Sigma_3$$

$$D = d_0 1 + d_1\Sigma_1 + d_2\Sigma_2 + d_3\Sigma_3.$$

The numbers b_0, b_1, b_2, b_3 and d_0, d_1, d_2, d_3 are real. Then

$$B + iD = z_0 1 + z_1\Sigma_1 + z_2\Sigma_2 + z_3\Sigma_3,$$

$$B - iD = z_0^* 1 + z_1^*\Sigma_1 + z_2^*\Sigma_2 + z_3^*\Sigma_3,$$

where

$$z_0 = b_0 + id_0,$$

$$z_1 = b_1 + id_1,$$

$$z_2 = b_2 + id_2,$$

$$z_3 = b_3 + id_3.$$

Therefore

$$\begin{aligned}(B + iD)(B - iD) = {} & (z_0z_0^* + z_1z_1^* + z_2z_2^* + z_3z_3^*)1 \\ & + (z_0z_1^* + z_0^*z_1 + iz_2z_3^* - iz_2^*z_3)\Sigma_1 \\ & + (z_0z_2^* + z_0^*z_2 + iz_3z_1^* - iz_3^*z_1)\Sigma_2 \\ & + (z_0z_3^* + z_0^*z_3 + iz_1z_2^* - iz_1^*z_2)\Sigma_3.\end{aligned}$$

This represents a real quantity, because the number in each ( ) is real. It represents a non-negative quantity, because

$$z_0z_0^* + z_1z_1^* + z_2z_2^* + z_3z_3^* \geq 0$$

and

$$\begin{aligned}(z_0z_0^* + z_1z_1^* + z_2z_2^* + z_3z_3^*)^2 \geq {} & (z_0z_1^* + z_0^*z_1 + iz_2z_3^* - iz_2^*z_3)^2 \\ & + (z_0z_2^* + z_0^*z_2 + iz_3z_1^* - iz_3^*z_1)^2 \\ & + (z_0z_3^* + z_0^*z_3 + iz_1z_2^* - iz_1^*z_2)^2.\end{aligned}$$

The latter follows from

$$\begin{aligned}
&(z_0 z_0^*)^2 + (z_1 z_1^*)^2 + (z_2 z_2^*)^2 + (z_3 z_3^*)^2 \\
&\quad \geq (z_0 z_1^*)^2 + (z_0^* z_1)^2 - (z_2 z_3^*)^2 - (z_2^* z_3)^2 \\
&\quad + (z_0 z_2^*)^2 + (z_0^* z_2)^2 - (z_3 z_1^*)^2 - (z_3^* z_1)^2 \\
&\quad + (z_0 z_3^*)^2 + (z_0^* z_3)^2 - (z_1 z_2^*)^2 - (z_1^* z_2)^2,
\end{aligned}$$

which follows from the fact that

$$(z_0 z_0 - z_1 z_1 - z_2 z_2 - z_3 z_3)(z_0^* z_0^* - z_1^* z_1^* - z_2^* z_2^* - z_3^* z_3^*) \geq 0.$$

PROBLEMS

11-1. Which of the matrices (a)–(f) of Problem 10-2 represent non-negative real physical quantities?

11-2. Show that for every 2×2 matrix K that represents a non-negative real quantity, there is a matrix L that represents a non-negative real quantity such that

$$K = L^2.$$

Do this by showing that the matrix L already considered represents a non-negative quantity.

12 WHAT CAN BE MEASURED

Now we find which quantities can have definite values together, for the same state. The values of these quantities are what can be measured. They determine the state. From them we compute probabilities for different possible values for other quantities. We still consider only quantities represented by 2×2 matrices that describe a spin and magnetic moment.

Again, let x_1, x_2, x_3 be real numbers, let

$$r = \sqrt{x_1^2 + x_2^2 + x_3^2},$$

and let

$$u_1 = \frac{x_1}{r}, \qquad u_2 = \frac{x_2}{r}, \qquad u_3 = \frac{x_3}{r},$$

so

$$u_1^2 + u_2^2 + u_3^2 = 1$$

and

$$x_1 = ru_1, \qquad x_2 = ru_2, \qquad x_3 = ru_3.$$

Again, consider the real quantities represented by the matrices

$$U = u_1\Sigma_1 + u_2\Sigma_2 + u_3\Sigma_3$$

and

$$A = x_1\Sigma_1 + x_2\Sigma_2 + x_3\Sigma_3 = rU.$$

The possible values of the quantity represented by U are 1 and -1, and the possible values of the quantity represented by A are r and $-r$.

Consider the mean values $\langle\Sigma_1\rangle, \langle\Sigma_2\rangle, \langle\Sigma_3\rangle$ for a given state. Which quantities have definite values for this state? We show that if any physical quantity has a definite value, then

$$\langle\Sigma_1\rangle^2 + \langle\Sigma_2\rangle^2 + \langle\Sigma_3\rangle^2 = 1.$$

We show that the quantity represented by A has the value r only if

$$u_1 = \langle\Sigma_1\rangle, \qquad u_2 = \langle\Sigma_2\rangle, \qquad u_3 = \langle\Sigma_3\rangle,$$

and we show it has the value $-r$ only if

$$u_1 = -\langle\Sigma_1\rangle, \qquad u_2 = -\langle\Sigma_2\rangle, \qquad u_3 = -\langle\Sigma_3\rangle.$$

Thus we find all the physical quantities that have definite values for a given state.

Suppose the quantity represented by A has the value r. Then $\langle A\rangle$ is r and $\langle U\rangle$ is 1, so

$$u_1\langle\Sigma_1\rangle + u_2\langle\Sigma_2\rangle + u_3\langle\Sigma_3\rangle = 1.$$

Therefore

$$\begin{aligned}
&(u_1 - \langle\Sigma_1\rangle)^2 + (u_2 - \langle\Sigma_2\rangle)^2 + (u_3 - \langle\Sigma_3\rangle)^2 \\
&\quad = u_1^2 + u_2^2 + u_3^2 + \langle\Sigma_1\rangle^2 + \langle\Sigma_2\rangle^2 + \langle\Sigma_3\rangle^2 \\
&\quad\quad - 2u_1\langle\Sigma_1\rangle - 2u_2\langle\Sigma_2\rangle - 2u_3\langle\Sigma_3\rangle \\
&\quad = 1 + \langle\Sigma_1\rangle^2 + \langle\Sigma_2\rangle^2 + \langle\Sigma_3\rangle^2 - 2 \\
&\quad = \langle\Sigma_1\rangle^2 + \langle\Sigma_2\rangle^2 + \langle\Sigma_3\rangle^2 - 1.
\end{aligned}$$

Since

$$(u_1 - \langle\Sigma_1\rangle)^2 + (u_2 - \langle\Sigma_2\rangle)^2 + (u_3 - \langle\Sigma_3\rangle)^2 \geq 0$$

and

$$\langle\Sigma_1\rangle^2 + \langle\Sigma_2\rangle^2 + \langle\Sigma_3\rangle^2 - 1 \leq 0,$$

these two quantities can be equal only if both are zero. This implies

$$\langle \Sigma_1 \rangle^2 + \langle \Sigma_2 \rangle^2 + \langle \Sigma_3 \rangle^2 = 1$$

and

$$u_1 = \langle \Sigma_1 \rangle, \qquad u_2 = \langle \Sigma_2 \rangle, \qquad u_3 = \langle \Sigma_3 \rangle.$$

Suppose the quantity represented by A has the value $-r$. Then $\langle A \rangle$ is $-r$ and $\langle U \rangle$ is -1, so

$$u_1\langle \Sigma_1 \rangle + u_2\langle \Sigma_2 \rangle + u_3\langle \Sigma_3 \rangle = -1.$$

Therefore

$$\begin{aligned}
&(u_1 + \langle \Sigma_1 \rangle)^2 + (u_2 + \langle \Sigma_2 \rangle)^2 + (u_3 + \langle \Sigma_3 \rangle)^2 \\
&= u_1^2 + u_2^2 + u_3^2 + \langle \Sigma_1 \rangle^2 + \langle \Sigma_2 \rangle^2 + \langle \Sigma_3 \rangle^2 \\
&\quad + 2u_1\langle \Sigma_1 \rangle + 2u_2\langle \Sigma_2 \rangle + 2u_3\langle \Sigma_3 \rangle \\
&= 1 + \langle \Sigma_1 \rangle^2 + \langle \Sigma_2 \rangle^2 + \langle \Sigma_3 \rangle^2 - 2 \\
&= \langle \Sigma_1 \rangle^2 + \langle \Sigma_2 \rangle^2 + \langle \Sigma_3 \rangle^2 - 1.
\end{aligned}$$

This implies

$$\langle \Sigma_1 \rangle^2 + \langle \Sigma_2 \rangle^2 + \langle \Sigma_3 \rangle^2 = 1$$

and

$$u_1 = -\langle \Sigma_1 \rangle, \qquad u_2 = -\langle \Sigma_2 \rangle, \qquad u_3 = -\langle \Sigma_3 \rangle.$$

For any matrix M that represents a physical quantity, there is a matrix A and numbers z_0 and c such that

$$M = z_0 1 + cA.$$

The quantity represented by M has a definite value if and only if the quantity represented by A has a definite value. Thus for a state that gives mean values $\langle \Sigma_1 \rangle, \langle \Sigma_2 \rangle, \langle \Sigma_3 \rangle$, there are physical quantities that have definite values only if

$$\langle \Sigma_1 \rangle^2 + \langle \Sigma_2 \rangle^2 + \langle \Sigma_3 \rangle^2 = 1.$$

Then every physical quantity that has a definite value is represented by a matrix

$$z_0 1 + cA,$$

where either

$$x_1 = r\langle\Sigma_1\rangle, \qquad x_2 = r\langle\Sigma_2\rangle, \qquad x_3 = r\langle\Sigma_3\rangle$$

or

$$x_1 = -r\langle\Sigma_1\rangle, \qquad x_2 = -r\langle\Sigma_2\rangle, \qquad x_3 = -r\langle\Sigma_3\rangle.$$

There is a definite value for the projection of the magnetic moment in the direction of the $\langle\Sigma_1\rangle, \langle\Sigma_2\rangle, \langle\Sigma_3\rangle$ vector. There is a corresponding definite value for every quantity obtained from that by multiplying by a number and adding a number. No other quantities have definite values.

Conversely, the state is determined if there is a definite value for the projection of the magnetic moment in some direction. Suppose the quantity represented by a particular matrix A has the value r. Then

$$\langle\Sigma_1\rangle = u_1, \qquad \langle\Sigma_2\rangle = u_2, \qquad \langle\Sigma_3\rangle = u_3.$$

These are determined by the given x_1, x_2, x_3 that specify A. From this information we can compute the mean values and probabilities for all physical quantities.

For example, suppose the quantity represented by the matrix

$$U = \tfrac{3}{5}\Sigma_1 + \tfrac{4}{5}\Sigma_2$$

has the value 1. This corresponds to the value μ for the projection of the magnetic moment in the direction specified by

$$u_1 = \tfrac{3}{5}, \qquad u_2 = \tfrac{4}{5}, \qquad u_3 = 0.$$

Then

$$\langle\Sigma_1\rangle = \tfrac{3}{5}, \qquad \langle\Sigma_2\rangle = \tfrac{4}{5}, \qquad \langle\Sigma_3\rangle = 0.$$

For the quantity represented by Σ_1, the probability that the value is 1 is $\frac{4}{5}$ and the probability that the value is -1 is $\frac{1}{5}$. For the quantity represented by Σ_2, the probability that the value is 1 is $\frac{9}{10}$ and the probability that the value is -1 is $\frac{1}{10}$. For the quantity represented by Σ_3, there are equal probabilities $\frac{1}{2}$ for the two values 1 and -1.

For any state only certain quantities have definite values. These values are all there is to know about the state. From them we can calculate the mean values and probabilities for all physical quantities. The quantities that have definite values are all that can be measured. If we measure a definite value for any other quantity, we have a different state.

We can measure the projection of the magnetic moment in one direction. If we obtain a definite value for it, we also have a definite value for every quantity we get by multiplying it by a number and adding a number. These are all the quantities we can measure. Projections of the magnetic moment in two different directions cannot be measured together. They cannot have definite values together for the same state.

How do we determine whether a pair of real quantities can have definite values together? Consider two real quantities represented by the matrices

$$K = x_0 1 + x_1 \Sigma_1 + x_2 \Sigma_2 + x_3 \Sigma_3$$

and

$$L = y_0 1 + y_1 \Sigma_1 + y_2 \Sigma_2 + y_3 \Sigma_3 .$$

It is easy to see that these quantities can have definite values together in certain cases, depending on the real numbers x_0, x_1, x_2, x_3 and y_0, y_1, y_2, y_3. If x_1, x_2, x_3 are 0, then

$$K = x_0 1 . \tag{i}$$

In this case, the quantity represented by K has the value x_0 for any state, so the two quantities have definite values together whenever the quantity represented by L has a definite value. If y_1, y_2, y_3 are 0, then

$$L = y_0 1 . \tag{ii}$$

In this case, the quantity represented by L has the value y_0 for any state, so the two quantities have definite values together whenever the quantity represented by K has a definite value. If there is a real number b such that

$$y_1 = bx_1, \qquad y_2 = bx_2, \qquad y_3 = bx_3, \tag{iii}$$

then

$$K = x_0 1 + A$$

and

$$L = y_0 1 + bA,$$

where

$$A = x_1 \Sigma_1 + x_2 \Sigma_2 + x_3 \Sigma_3 .$$

In this case, the quantities represented by K and L both have definite values whenever the quantity represented by A has a definite value.

These are the only ways two real quantities can have definite values together. There are no other cases of the real numbers x_0, x_1, x_2, x_3 and y_0, y_1, y_2, y_3 that allow the two quantities to have definite values for the same state. If the quantity represented by K has a definite value for a state that gives mean values $\langle\Sigma_1\rangle, \langle\Sigma_2\rangle, \langle\Sigma_3\rangle$, then there is a real number x (which is either r or $-r$) such that

$$x_1 = x\langle\Sigma_1\rangle, \qquad x_2 = x\langle\Sigma_2\rangle, \qquad x_3 = x\langle\Sigma_3\rangle.$$

If the quantity represented by L has a definite value for the same state, there is also a real number y such that

$$y_1 = y\langle\Sigma_1\rangle, \qquad y_2 = y\langle\Sigma_2\rangle, \qquad y_3 = y\langle\Sigma_3\rangle.$$

If x is 0, we have case (i). If x is not 0, we have case (iii) with

$$b = \frac{y}{x}.$$

These are the same as the cases in which the two matrices K and L commute. Two real quantities represented by matrices K and L have definite values together, for the same state, if and only if

$$KL = LK.$$

In every case, both matrices K and L are obtained from a single matrix A that represents a real quantity, so if the quantity represented by A has a definite value, then the quantities represented by both K and L have corresponding definite values. The quantities represented by K and L have definite values together in every state where the quantity represented by A has a definite value. In case (iii) we have

$$K = x_0 1 + A$$

and

$$L = y_0 1 + bA$$

with

$$A = x_1\Sigma_1 + x_2\Sigma_2 + x_3\Sigma_3.$$

In case (ii) we have the same with 0 for b. In case (i) we have

$$K = x_0 1 + 0 \cdot A$$

and

$$L = y_0 1 + A$$

with

$$A = y_1 \Sigma_1 + y_2 \Sigma_2 + y_3 \Sigma_3 .$$

This is possible only if K and L commute, because all the matrices

$$(\text{number})1 + (\text{number})A$$

commute with each other.

If the two matrices K and L represent real quantities that have definite values together, then the matrix

$$M = K + iL$$

has the form

$$M = z_0 1 + cA$$

that we found previously for a complex physical quantity. For case (iii) we have

$$K + iL = (x_0 + iy_0)1 + (1 + ib)(x_1 \Sigma_1 + x_2 \Sigma_2 + x_3 \Sigma_3)$$

$$= z_0 1 + cA$$

with

$$z_0 = x_0 + iy_0,$$

$$c = 1 + ib,$$

and

$$A = x_1 \Sigma_1 + x_2 \Sigma_2 + x_3 \Sigma_3 .$$

Case (ii) is the same with 0 for b. For case (i) we have

$$K + iL = x_0 1 + i(y_0 1 + y_1 \Sigma_1 + y_2 \Sigma_2 + y_3 \Sigma_3),$$

which has the form

$$z_0 1 + cA$$

with

$$z_0 = x_0 + iy_0,$$

$$c = i,$$

and

$$A = y_1\Sigma_1 + y_2\Sigma_2 + y_3\Sigma_3.$$

Conversely, if M is a matrix that represents a complex physical quantity, there are matrices K and L such that

$$M = K + iL$$

and K and L represent real quantities that have definite values together. Given

$$M = z_0 1 + cA$$

in terms of complex numbers z_0 and c, let

$$z_0 = x_0 + iy_0,$$

$$c = d + if,$$

and

$$A = a_1\Sigma_1 + a_2\Sigma_2 + a_3\Sigma_3$$

in terms of real numbers x_0, y_0, d, f and a_1, a_2, a_3. Then

$$M = (x_0 + iy_0)1 + (d + if)(a_1\Sigma_1 + a_2\Sigma_2 + a_3\Sigma_3)$$

$$= K + iL,$$

where K is specified by x_0 and

$$x_1 = da_1, \qquad x_2 = da_2, \qquad x_3 = da_3$$

and L is specified by y_0 and

$$y_1 = fa_1, \qquad y_2 = fa_2, \qquad y_3 = fa_3.$$

If d is 0, we have case (i). If d is not 0, we have case (iii) with

$$b = \frac{f}{d}.$$

Thus a matrix M represents a physical quantity if and only if

$$M = K + iL,$$

where K and L are matrices representing real quantities that have definite values together. Then M represents a complex quantity made from real and imaginary parts that have definite values together.

PROBLEMS

12-1. Consider the state where the quantity represented by $4\Sigma_2 + 3\Sigma_3$ has a definite and positive value. For the quantity represented by each of the following matrices:

(i) find the mean value for this state;
(ii) determine whether the quantity has a definite value for this state; and, if it does,
(iii) find the value.

(a) $4\Sigma_2 + 3\Sigma_3$
(b) Σ_1
(c) Σ_2
(d) Σ_3
(e) $3\Sigma_2 + 4\Sigma_3$
(f) $4\Sigma_2 - 3\Sigma_3$
(g) $3\Sigma_1 + 4\Sigma_2$
(h) $3\Sigma_1 + 4\Sigma_2 + 3\Sigma_3$
(i) $8\Sigma_2 + 6\Sigma_3$
(j) $(5)1 + 4\Sigma_2 + 3\Sigma_3$

12-2. Which pairs of the following matrices represent real quantities that can have definite values together?

(a) Σ_1
(b) Σ_2
(c) Σ_3
(d) $\Sigma_1 + \Sigma_2$
(e) $\Sigma_1 - \Sigma_2$
(f) $-\Sigma_1 + \Sigma_2$.

12-3. Suppose you know that

$$2\langle\Sigma_1\rangle + \langle\Sigma_2\rangle + 2\langle\Sigma_3\rangle = 3.$$

Is that enough to determine $\langle\Sigma_1\rangle, \langle\Sigma_2\rangle, \langle\Sigma_3\rangle$? If it is, find them. If it is not, explain why not.

12-4. Let K and L be matrices that represent real quantities. Suppose

$$K = L^2.$$

Show that K and L commute. This implies there is a matrix

$$A = x_1\Sigma_1 + x_2\Sigma_2 + x_3\Sigma_3,$$

where x_1, x_2, x_3 are real numbers, such that

$$K = x_0 1 + gA$$

$$L = y_0 1 + bA$$

with real numbers x_0, y_0, g, b. Show that

$$x_0 = y_0{}^2 + b^2 r^2$$

$$g = 2y_0 b,$$

where

$$r^2 = x_1{}^2 + x_2{}^2 + x_3{}^2.$$

Suppose L represents a non-negative quantity. Show that then there is only one solution for y_0 and b in terms of x_0, g and r^2. There is only one matrix L that represents a non-negative real quantity such that L^2 is K. It was shown in Problem 11-2 that there is one such matrix L for each 2×2 matrix K that represents a non-negative real quantity. We can call it $\sqrt{K}$.

Show that if M is a matrix that commutes with K, then M commutes with A, and therefore M commutes with L. Every matrix that commutes with K also commutes with $\sqrt{K}$.

13 GENERAL RULES

The description of a magnetic moment in terms of 2×2 matrices is a particularly simple example of quantum mechanics. Some of the properties we found are more characteristic of this example than of quantum mechanics in general. However, we have learned some things that are generally true in quantum mechanics. We list them here and develop some of their implications for later use.

Certain matrices represent real quantities. The definition of a real quantity is that the possible values are real numbers. Therefore the mean value of a real quantity is a real number for any state. Conversely, if a quantity is represented by a matrix M such that $\langle M \rangle$ is real for every state, then it is a real quantity; each possible value is real, because $\langle M \rangle$ is that value for some state. For each possible value that is not part of a continuous range of possible values, there is a state where the quantity has that definite value. For each possible value that is part of a continuous range, there is a state where the probabilities give that value for the mean.

If K is a matrix that represents a real quantity and b is a real number, then the matrix bK represents a real quantity; for any state, its mean value

$$\langle bK \rangle = b\langle K \rangle$$

is a real number, because $\langle K \rangle$ is a real number.

If K and L are matrices that represent real quantities, then the matrix $K + L$ represents a real quantity. That is true in general as well as in the 2×2 example. In general, we can see that if $K + L$ represents any quantity, it must be a real quantity; for any state, its mean value

$$\langle K + L \rangle = \langle K \rangle + \langle L \rangle$$

is a real number, because $\langle K \rangle$ and $\langle L \rangle$ are real numbers.

For any state, some quantities have definite values and others do not. In particular, there is no state where all quantities have definite values. There is always some quantity that does not have a definite value. This fact is often called von Neumann's theorem on hidden variables [1]. We have proved it for the 2×2 example, and our proof can be extended easily to apply to quantum mechanics in general.‡

Real quantities represented by matrices that commute have definite values together. Again, let K and L be matrices that represent real quantities. If K and L commute, both can be obtained from a single matrix A that represents a real quantity, so that if the quantity represented by A has a definite value, then the quantities represented by both K and L have corresponding definite values. Then the quantities represented by K and L have definite values together in every state where the quantity represented by A has a definite value. This is not possible if K and L do not commute, but then there may be some states where both quantities represented by K and L have definite values.

A real quantity is called non-negative if all its possible values are non-negative real numbers. Then its mean value is a non-negative real number for any state. Conversely, if a real quantity is represented by a matrix M such that $\langle M \rangle$ is a non-negative real number for any state, then it is a non-negative quantity; each possible value is a non-negative real number, because $\langle M \rangle$ is that value for some state. To indicate that the matrix M represents a non-negative real quantity, we write

$$M \geq 0.$$

If L is a matrix that represents a real quantity, the matrix L^2 represents a non-negative real quantity, so

$$\langle L^2 \rangle \geq 0$$

for any state. Conversely, if K is a matrix that represents a non-negative real quantity, there is a matrix L that represents a real quantity such that

$$K = L^2.$$

Let B and D be matrices that represent real quantities. Then the matrix

$$(B + iD)(B - iD)$$

‡For any system and any state vector, there are matrices that are the same as the Pauli matrices for a two-dimensional subspace containing the state vector and zero otherwise.

represents a non-negative real quantity, so

$$\langle (B + iD)(B - iD) \rangle \geq 0$$

for any state. If the matrices B and D commute, so the quantities represented by B and D have definite values together, then the matrix $B + iD$ represents the complex quantity whose real and imaginary parts are represented by B and D. Then the square of the absolute value of this complex quantity is represented by the matrix

$$\begin{aligned}(B + iD)(B - iD) &= B^2 + D^2 - iBD + iDB \\ &= B^2 + D^2.\end{aligned}$$

Thus we can see that this matrix represents a non-negative real quantity when B and D commute. It is still true when B and D do not commute.

All these rules, which were shown to be true for the 2×2 example, are also true for quantum mechanics in general. Now we develop some implications of these rules for later use.

Again, let B and D be matrices that represent real quantities. The rules imply that the matrix

$$BD + DB = (B + D)^2 - B^2 - D^2$$

represents a real quantity, so the mean value

$$\langle BD + DB \rangle = \langle BD \rangle + \langle DB \rangle$$

is real, for any state. The rules also imply that the matrix

$$-i(BD - DB) = (B + iD)(B - iD) - B^2 - D^2$$

represents a real quantity, so

$$-i\langle BD - DB \rangle$$

is real and

$$\langle BD - DB \rangle = \langle BD \rangle - \langle DB \rangle$$

is imaginary, for any state. These imply that the real parts of $\langle BD \rangle$ and $\langle DB \rangle$ are the same and the imaginary parts are opposite, which means

$$\langle DB \rangle = \langle BD \rangle^*,$$

for any state. To see this, let

$$\langle BD \rangle = x + iy$$

and

$$\langle DB \rangle = u + iv$$

in terms of real numbers x, y and u, v. Then

$$x + iy + u + iv$$

is real, which implies

$$v = -y,$$

and

$$x + iy - (u + iv)$$

is imaginary, which implies

$$u = x.$$

Therefore

$$u + iv = x - iy = (x + iy)^*$$

so

$$\langle DB \rangle = \langle BD \rangle^*.$$

In particular, suppose two real quantities are represented by matrices B and D that commute. Then the matrix

$$BD = DB = \tfrac{1}{2}(BD + DB)$$

represents a real quantity, and the mean value

$$\langle BD \rangle = \langle DB \rangle = \langle BD \rangle^*$$

is a real number. Since the quantities represented by B and D have definite values together, it is meaningful to consider their sum. It has a definite value when they do; its value is the sum of their values. It is measured when they are. For any state its mean value is

$$\langle B \rangle + \langle D \rangle.$$

That implies it is represented by the matrix $B + D$. To see this, consider the quantity represented by $B + D$. Since $B + D$ commutes with both B and D, it is meaningful to consider the quantity you get by subtracting the quantity represented by $B + D$ from the sum of the quantities represented by B and D. It has the mean value

$$\langle B \rangle + \langle D \rangle - \langle B + D \rangle = 0$$

for any state, so it can have only the value 0. Thus, there is no difference between the sum of the quantities represented by B and D and the quantity represented by $B + D$. The sum of the quantities represented by the matrices B and D is represented by the matrix $B + D$. The square of the sum is represented by the matrix

$$(B + D)^2 = B^2 + D^2 + 2BD.$$

Then the matrix

$$BD = \tfrac{1}{2}\left[(B + D)^2 - B^2 - D^2\right]$$

represents the product of the quantities represented by B and D, because that is half the quantity you get by subtracting the squares of the quantities represented by B and D from the square of their sum. The product of the quantities represented by B and D has a definite value when the quantities represented by B and D do; its value is the product of their values. It is measured when they are.

Let K and L be matrices that represent real quantities, and let w be a complex number. The rules imply that

$$(K + wL)(K + w^*L) \geq 0.$$

To see this, let

$$w = u + iv$$

in terms of real numbers u and v, and let

$$B = K + uL$$

and

$$D = vL.$$

Then B and D represent real quantities, and

$$K + wL = K + uL + ivL$$

$$= B + iD$$

and

$$\begin{aligned} K + w^*L &= K + uL - ivL \\ &= B - iD, \end{aligned}$$

so

$$(K + wL)(K + w^*L) = (B + iD)(B - iD) \geq 0.$$

Let K and L be matrices that represent real quantities, and let G be a matrix that represents a non-negative real quantity. The rules imply that

$$(K + iL)G(K - iL) \geq 0.$$

To see this, let R be a matrix that represents a real quantity such that

$$G = R^2$$

and let

$$\begin{aligned} B &= \tfrac{1}{2}(KR + RK) + \tfrac{1}{2}i(LR - RL) \\ D &= \tfrac{1}{2}(LR + RL) - \tfrac{1}{2}i(KR - RK). \end{aligned}$$

Then B and D represent real quantities, and

$$\begin{aligned} (K + iL)G(K - iL) &= (K + iL)RR(K - iL) \\ &= \tfrac{1}{2}[(KR + RK) + i(LR - RL) + (KR - RK) \\ &\qquad + i(LR + RL)] \\ &\quad \times \tfrac{1}{2}[(RK + KR) - i(RL - LR) \\ &\qquad + (RK - KR) - i(RL + LR)] \\ &= (B + iD)(B - iD) \geq 0. \end{aligned}$$

REFERENCES

1. J. von Neumann, *Mathematical Foundations of Quantum Mechanics*, Princeton University Press, 1955, IV, 2.

14 TWO SPINS

Now we consider two spins, and the corresponding two magnetic moments, for a system of two particles, so we can discuss experiments done with them. The significance of these experiments is explained in the following two chapters.

For one spin we use matrices $\Sigma_1, \Sigma_2, \Sigma_3$ with the same multiplication rules as before. For the other spin we use matrices Ξ_1, Ξ_2, Ξ_3 with the same multiplication rules as $\Sigma_1, \Sigma_2, \Sigma_3$. Thus

$$\Xi_1{}^2 = 1, \qquad \Xi_2{}^2 = 1, \qquad \Xi_3{}^2 = 1,$$

$$\Xi_1\Xi_2 = i\Xi_3, \qquad \Xi_2\Xi_1 = -i\Xi_3,$$

$$\Xi_2\Xi_3 = i\Xi_1, \qquad \Xi_3\Xi_2 = -i\Xi_1,$$

$$\Xi_3\Xi_1 = i\Xi_2, \qquad \Xi_1\Xi_3 = -i\Xi_2.$$

Each Σ matrix commutes with each Ξ matrix. Thus

$$\Sigma_j\Xi_k = \Xi_k\Sigma_j$$

for $j = 1, 2, 3$ and $k = 1, 2, 3$. This means projections of both spins have definite values together. They can be measured together because they are for different particles.

All these cannot be 2×2 matrices. We showed that if a 2×2 matrix commutes with the 2×2 Pauli matrices $\Sigma_1, \Sigma_2, \Sigma_3$, then it is the matrix 1 multiplied by a number. None of the matrices Ξ_1, Ξ_2, Ξ_3 can be the matrix 1 multiplied by a number, because they do not commute with each other. There are 4×4 matrices for $\Sigma_1, \Sigma_2, \Sigma_3$ and Ξ_1, Ξ_2, Ξ_3. They are described in Problem 14-4, to satisfy curiosity. We do not use them. We calculate everything we want algebraically, using only the multiplication rules.

We want to discuss experiments done with two particles in a state where the total spin is zero. The total spin is the vector quantity whose projections in the three perpendicular reference directions are represented by the matrices

$$\tfrac{1}{2}\hbar(\Sigma_1 + \Xi_1), \qquad \tfrac{1}{2}\hbar(\Sigma_2 + \Xi_2), \qquad \tfrac{1}{2}\hbar(\Sigma_3 + \Xi_3).$$

When we say the total spin is zero, we mean that for any real numbers x_1, x_2, x_3, the quantity represented by the matrix

$$x_1(\Sigma_1 + \Xi_1) + x_2(\Sigma_2 + \Xi_2) + x_3(\Sigma_3 + \Xi_3)$$

has the value 0. The projection of the total spin in any direction is zero. In particular, the quantity represented by $\Sigma_1 + \Xi_1$ has the value 0. Therefore

$$\left\langle (\Sigma_1 + \Xi_1)^2 \right\rangle = 0.$$

Since

$$\begin{aligned}(\Sigma_1 + \Xi_1)^2 &= \Sigma_1{}^2 + \Xi_1{}^2 + \Sigma_1\Xi_1 + \Xi_1\Sigma_1 \\ &= 2 + 2\Sigma_1\Xi_1,\end{aligned}$$

it follows that

$$\langle \Sigma_1\Xi_1 \rangle = -1.$$

It follows similarly that

$$\langle \Sigma_2\Xi_2 \rangle = -1$$

and

$$\langle \Sigma_3\Xi_3 \rangle = -1.$$

We get more information about this state by using the simple algebraic properties of the matrices. Let U and V be matrices that represent real quantities, such that

$$U^2 = 1, \qquad V^2 = 1,$$

and

$$UV + VU = 0.$$

Then, for any state,

$$\langle U\rangle^2 + \langle V\rangle^2 \leq 1.$$

We prove this the same way we proved that

$$\langle \Sigma_1\rangle^2 + \langle \Sigma_2\rangle^2 + \langle \Sigma_3\rangle^2 \leq 1.$$

Let x and y be real numbers, and let

$$r = \sqrt{x^2 + y^2}.$$

The matrix

$$A = xU + yV$$

represents a real quantity. The square of this quantity is represented by

$$(xU + yV)^2 = (x^2 + y^2)1 = r^2 1,$$

so it can have only the value r^2. Therefore the quantity represented by A can have only the values $-r$ and r. For any state its mean value is a real number between $-r$ and r. Thus

$$-r \leq x\langle U\rangle + y\langle V\rangle \leq r.$$

This holds for any real numbers x, y. In particular, let

$$x = \langle U\rangle, \qquad y = \langle V\rangle.$$

Then

$$xx + yy \leq r$$

or

$$r^2 \leq r.$$

This implies

$$r \leq 1$$

and

$$r^2 \leq 1$$

or

$$\langle U \rangle^2 + \langle V \rangle^2 \leq 1.$$

The general rules imply that the matrices

$$\Sigma_1 \Xi_1 = \Xi_1 \Sigma_1$$

and

$$\Sigma_2 \Xi_1 = \Xi_1 \Sigma_2$$

represent real quantities. We have

$$(\Sigma_1 \Xi_1)^2 = \Sigma_1{}^2 \Xi_1{}^2 = 1,$$

$$(\Sigma_2 \Xi_1)^2 = \Sigma_2{}^2 \Xi_1{}^2 = 1,$$

and

$$\Sigma_1 \Xi_1 \Sigma_2 \Xi_1 + \Sigma_2 \Xi_1 \Sigma_1 \Xi_1 = \Sigma_1 \Sigma_2 \Xi_1{}^2 + \Sigma_2 \Sigma_1 \Xi_1{}^2 = 0.$$

Therefore, letting

$$U = \Sigma_1 \Xi_1, \qquad V = \Sigma_2 \Xi_1,$$

we get

$$\langle \Sigma_1 \Xi_1 \rangle^2 + \langle \Sigma_2 \Xi_1 \rangle^2 \leq 1$$

for any state. For a state where the total spin is zero, we have

$$\langle \Sigma_1 \Xi_1 \rangle = -1.$$

Then

$$\langle \Sigma_2 \Xi_1 \rangle = 0.$$

We can show similarly that

$$\langle \Sigma_1 \Xi_2 \rangle = 0,$$

$$\langle \Sigma_2 \Xi_3 \rangle = 0, \qquad \langle \Sigma_3 \Xi_2 \rangle = 0,$$

$$\langle \Sigma_3 \Xi_1 \rangle = 0, \qquad \langle \Sigma_1 \Xi_3 \rangle = 0$$

for a state where the total spin is zero. Let a_1, a_2, a_3 and b_1, b_2, b_3 be real numbers. Then

$$\langle (a_1\Sigma_1 + a_2\Sigma_2 + a_3\Sigma_3)(b_1\Xi_1 + b_2\Xi_2 + b_3\Xi_3) \rangle = -a_1b_1 - a_2b_2 - a_3b_3$$

for a state where the total spin is zero.

We also have

$$(\Sigma_1\Xi_1)^2 = 1, \qquad \Sigma_2{}^2 = 1,$$

$$\Sigma_1\Xi_1\Sigma_2 + \Sigma_2\Sigma_1\Xi_1 = 0,$$

so, letting

$$U = \Sigma_1\Xi_1, \qquad V = \Sigma_2,$$

we get

$$\langle \Sigma_1\Xi_1 \rangle^2 + \langle \Sigma_2 \rangle^2 \leq 1.$$

For a state where the total spin is zero, it follows that

$$\langle \Sigma_2 \rangle = 0.$$

We can show similarly that

$$\langle \Sigma_1 \rangle = 0, \qquad \langle \Sigma_3 \rangle = 0,$$

$$\langle \Xi_1 \rangle = 0, \qquad \langle \Xi_2 \rangle = 0, \qquad \langle \Xi_3 \rangle = 0$$

for a state where the total spin is zero.

These quantities are measured in experiments where a reaction produces a pair of particles, such as protons, in a state where the total spin is zero. The particles come out in opposite directions. After they are separated, a projection of the magnetic moment in some direction is measured for each of the two particles.

Let a_1, a_2, a_3 be real numbers such that

$$a_1^2 + a_2^2 + a_3^2 = 1.$$

Measuring the projection of the magnetic moment in the direction of the a_1, a_2, a_3 vector for one of the particles determines a value, either 1 or -1, for the quantity represented by

$$a_1\Sigma_1 + a_2\Sigma_2 + a_3\Sigma_3.$$

Measuring the projection of the magnetic moment in that direction for the other particle determines a value, either 1 or -1, for the quantity represented by

$$a_1\Xi_1 + a_2\Xi_2 + a_3\Xi_3.$$

The fact that the total spin is 0 means the sum of these quantities has the value 0. That implies the two quantities have opposite values, either 1 and -1 or -1 and 1, but not both 1 and not both -1. By observing that this is true for every pair of particles put through such measurements, for different a_1, a_2, a_3 direction vectors, the experimenter can verify that the reaction does always produce a pair of particles in a state where the total spin is zero.

For this state we have

$$\langle a_1\Sigma_1 + a_2\Sigma_2 + a_3\Sigma_3\rangle = a_1\langle\Sigma_1\rangle + a_2\langle\Sigma_2\rangle + a_3\langle\Sigma_3\rangle = 0$$

so there are equal probabilities $\frac{1}{2}$ for the two possible values 1 and -1 of the quantity represented by

$$a_1\Sigma_1 + a_2\Sigma_2 + a_3\Sigma_3.$$

We also have

$$\langle a_1\Xi_1 + a_2\Xi_2 + a_3\Xi_3\rangle = a_1\langle\Xi_1\rangle + a_2\langle\Xi_2\rangle + a_3\langle\Xi_3\rangle = 0$$

so there are equal probabilities $\frac{1}{2}$ for the two possible values 1 and -1 of the quantity represented by

$$a_1\Xi_1 + a_2\Xi_2 + a_3\Xi_3.$$

This implies that the probability of finding the pair of values 1, -1 for the two quantities is the same as the probability of finding the pair of values $-1, 1$. Both probabilities are $\frac{1}{2}$. The probability of the pair of values $1, 1$ is 0, and the probability of the pair of values $-1, -1$ is 0. Thus we obtain a table of probabilities for the four pairs of values.

$1, 1 \;:\; 0$	$1, -1 \;:\; \frac{1}{2}$
$-1, 1 \;:\; \frac{1}{2}$	$-1, -1 \;:\; 0$

Projections of the two magnetic moments in different directions also are measured. Let b_1, b_2, b_3 also be real numbers such that

$$b_1^2 + b_2^2 + b_3^2 = 1.$$

Measuring the projection of the magnetic moment of one particle in the direction of the a_1, a_2, a_3 vector and the projection of the magnetic moment of the other particle in the direction of the b_1, b_2, b_3 vector determines a value, either 1 or -1, for each of the two quantities represented by

$$a_1\Sigma_1 + a_2\Sigma_2 + a_3\Sigma_3$$

and

$$b_1\Xi_1 + b_2\Xi_2 + b_3\Xi_3.$$

In general, there are nonzero probabilities for all four possible pairs of values

$1, \; 1$	$1, \; -1$
$-1, \; 1$	$-1, \; -1$

It is easy to calculate these probabilities. We still assume the total spin is zero. As before, the mean value of each quantity is 0, so for each quantity there are equal probabilities $\frac{1}{2}$ for the two possible values 1 and -1. We get the additional information we need by considering the product of the two quantities. It is represented by the product of the two matrices. We already showed that its mean value is

$$-a_1b_1 - a_2b_2 - a_3b_3.$$

The product of the two quantities can have only the values 1 and -1. It is 1 when the pair of values of the two quantities is $1,1$ or $-1,-1$, and it is -1 when the pair is $1,-1$ or $-1,1$. There is a probability $\rho(1)$ that the value of the product is 1 and a probability $\rho(-1)$ that the value of the product is -1, so that

$$\rho(1) + \rho(-1) = 1$$

$$(1)\rho(1) + (-1)\rho(-1) = -a_1 b_1 - a_2 b_2 - a_3 b_3.$$

By adding and subtracting these equations we get

$$\rho(1) = \tfrac{1}{2} + \tfrac{1}{2}(-a_1 b_1 - a_2 b_2 - a_3 b_3),$$

$$\rho(-1) = \tfrac{1}{2} - \tfrac{1}{2}(-a_1 b_1 - a_2 b_2 - a_3 b_3).$$

From this information we obtain the probabilities for the four possible pairs of values. They are shown in the table.

$1,1$: $\frac{1}{2}\rho(1)$	$1,-1$: $\frac{1}{2}\rho(-1)$
$-1,1$: $\frac{1}{2}\rho(-1)$	$-1,-1$: $\frac{1}{2}\rho(1)$

This is the only way to assign probabilities μ to the four pairs of possible values so that

$$\mu(1,1) + \mu(-1,-1) = \rho(1)$$

$$\mu(1,-1) + \mu(-1,1) = \rho(-1)$$

$$\mu(1,1) + \mu(1,-1) = \tfrac{1}{2}$$

$$\mu(-1,1) + \mu(-1,-1) = \tfrac{1}{2}$$

$$\mu(1,1) + \mu(-1,1) = \tfrac{1}{2}$$

$$\mu(1,-1) + \mu(-1,-1) = \tfrac{1}{2}.$$

We shall discuss some particular examples. We use

$$a_1 = 0, \qquad a_2 = 1, \qquad a_3 = 0$$

or

$$\alpha_1 = \tfrac{3}{5}, \qquad \alpha_2 = -\tfrac{4}{5}, \qquad \alpha_3 = 0$$

in place of a_1, a_2, a_3. We combine these two alternatives for a_1, a_2, a_3 with two alternatives for b_1, b_2, b_3, either

$$b_1 = 0, \qquad b_2 = 1, \qquad b_3 = 0$$

or

$$\beta_1 = 1, \qquad \beta_2 = 0, \qquad \beta_3 = 0$$

in place of b_1, b_2, b_3. That gives us four alternative combinations for a_1, a_2, a_3 and b_1, b_2, b_3, which we call the ab, $a\beta$, αb, $\alpha\beta$ combinations. The ab combination is an example of the case we considered first, where a_1, a_2, a_3 and b_1, b_2, b_3 are the same. For that we have numbers (0 and $\frac{1}{2}$) in the table of probabilities for the four possible pairs of values. Now we calculate numbers for the table of probabilities for each of the other three combinations. Since

$$-a_1\beta_1 - a_2\beta_2 - a_3\beta_3 = 0,$$

we get

$$\rho(1) = \tfrac{1}{2}$$

and

$$\rho(-1) = \tfrac{1}{2}$$

for the $a\beta$ combination. Since

$$-\alpha_1 b_1 - \alpha_2 b_2 - \alpha_3 b_3 = \tfrac{4}{5},$$

we get

$$\rho(1) = \tfrac{9}{10}$$

and

$$\rho(-1) = \tfrac{1}{10}$$

for the ab combination. Since

$$-\alpha_1\beta_1 - \alpha_2\beta_2 - \alpha_3\beta_3 = -\tfrac{3}{5},$$

we get

$$\rho(1) = \tfrac{1}{5}$$

and

$$\rho(-1) = \tfrac{4}{5}$$

for the $\alpha\beta$ combination. For each combination the probabilities for the four possible pairs of values are shown in the table. (The arrows between tables are for later use.)

ab	
$1, 1 : 0$	$1, -1 : \frac{1}{2}$
$-1, 1 : \frac{1}{2}$	$-1, -1 : 0$

$\alpha\beta$	
$1, 1 : \frac{1}{4}$	$1, -1 : \frac{1}{4}$
$-1, 1 : \frac{1}{4}$	$-1, -1 : \frac{1}{4}$

ab	
$1, 1 : \frac{9}{20}$	$1, -1 : \frac{1}{20}$
$-1, 1 : \frac{1}{20}$	$-1, -1 : \frac{9}{20}$

$\alpha\beta$	
$1, 1 : \frac{1}{10}$	$1, -1 : \frac{4}{10}$
$-1, 1 : \frac{4}{10}$	$-1, -1 : \frac{1}{10}$

We use one other example. Let

$$\gamma_1 = 0, \qquad \gamma_2 = 1, \qquad \gamma_3 = 0$$

$$\delta_1 = \frac{\sqrt{3}}{2}, \qquad \delta_2 = -\tfrac{1}{2}, \qquad \delta_3 = 0$$

$$\varepsilon_1 = -\frac{\sqrt{3}}{2}, \qquad \varepsilon_2 = -\tfrac{1}{2}, \qquad \varepsilon_3 = 0.$$

Then

$$-\gamma_1\delta_1 - \gamma_2\delta_2 - \gamma_3\delta_3 = \tfrac{1}{2}$$

$$-\delta_1\varepsilon_1 - \delta_2\varepsilon_2 - \delta_3\varepsilon_3 = \tfrac{3}{4} - \tfrac{1}{4} = \tfrac{1}{2}$$

$$-\varepsilon_1\gamma_1 - \varepsilon_2\gamma_2 - \varepsilon_3\gamma_3 = \tfrac{1}{2}.$$

Thus if any one of these vectors is used for a_1, a_2, a_3 and any other for b_1, b_2, b_3, provided the two are different, we get

$$\rho(-1) = \tfrac{1}{4}.$$

The probability that the two values in the pair are different, either $1, -1$ or $-1, 1$, is $\frac{1}{4}$.

PROBLEMS

14-1. Show that for any state

$$\langle\Sigma_1\Xi_1\rangle^2 + \langle\Sigma_3\rangle^2 \leq 1,$$
$$\langle\Sigma_1\Xi_2\rangle^2 + \langle\Sigma_3\rangle^2 \leq 1,$$
$$\langle\Sigma_1\Xi_3\rangle^2 + \langle\Sigma_3\rangle^2 \leq 1,$$
$$\langle\Sigma_2\Xi_1\rangle^2 + \langle\Sigma_3\rangle^2 \leq 1,$$
$$\langle\Sigma_2\Xi_2\rangle^2 + \langle\Sigma_3\rangle^2 \leq 1,$$
$$\langle\Sigma_2\Xi_3\rangle^2 + \langle\Sigma_3\rangle^2 \leq 1,$$
$$\langle\Sigma_3\Xi_1\rangle^2 + \langle\Xi_3\rangle^2 \leq 1,$$
$$\langle\Sigma_3\Xi_2\rangle^2 + \langle\Xi_3\rangle^2 \leq 1.$$

14-2. Use the result of the preceding problem to find the mean values

$$\langle\Sigma_1\rangle,\quad \langle\Sigma_2\rangle,\quad \langle\Sigma_3\rangle,\quad \langle\Xi_1\rangle,\quad \langle\Xi_2\rangle,\quad \langle\Xi_3\rangle,$$

$$\langle\Sigma_1\Xi_1\rangle,\quad \langle\Sigma_1\Xi_2\rangle,\quad \langle\Sigma_1\Xi_3\rangle,\quad \langle\Sigma_2\Xi_1\rangle,\quad \langle\Sigma_2\Xi_2\rangle,$$

$$\langle\Sigma_2\Xi_3\rangle,\quad \langle\Sigma_3\Xi_1\rangle,\quad \langle\Sigma_3\Xi_2\rangle,\quad \langle\Sigma_3\Xi_3\rangle$$

for a state where the quantities represented by Σ_3 and Ξ_3 both have the value 1.

14-3. Consider the two quantities represented by

$$a_1\Sigma_1 + a_2\Sigma_2 + a_3\Sigma_3$$

and

$$b_1\Xi_1 + b_2\Xi_2 + b_3\Xi_3$$

with

$$a_1 = 0, \qquad a_2 = \frac{\sqrt{3}}{2}, \qquad a_3 = -\tfrac{1}{2}$$

and

$$b_1 = 0, \qquad b_2 = -\frac{\sqrt{3}}{2}, \qquad b_3 = -\tfrac{1}{2}.$$

Find the mean value for each quantity, the probabilities for the possible values 1 and -1 for each quantity, the probability that the two quantities have the same value, the probability they have different values, and the probabilities for the four possible pairs of values

1, 1	1, -1
-1, 1	-1, -1

all for the state where the quantities represented by Σ_3 and Ξ_3 both have the value 1. Show the probability for each pair of values is the product of the probabilities for the two values. Do not assume that.

14-4. Let

$$\Sigma_1 = \begin{pmatrix} 0 & 0 & 1 & 0 \\ 0 & 0 & 0 & 1 \\ 1 & 0 & 0 & 0 \\ 0 & 1 & 0 & 0 \end{pmatrix}, \qquad \Sigma_2 = \begin{pmatrix} 0 & 0 & -i & 0 \\ 0 & 0 & 0 & -i \\ i & 0 & 0 & 0 \\ 0 & i & 0 & 0 \end{pmatrix},$$

$$\Xi_1 = \begin{pmatrix} 0 & 1 & 0 & 0 \\ 1 & 0 & 0 & 0 \\ 0 & 0 & 0 & 1 \\ 0 & 0 & 1 & 0 \end{pmatrix}, \qquad \Xi_2 = \begin{pmatrix} 0 & -i & 0 & 0 \\ i & 0 & 0 & 0 \\ 0 & 0 & 0 & -i \\ 0 & 0 & i & 0 \end{pmatrix}.$$

Calculate Σ_3 and Ξ_3. Show that these matrices satisfy all the multiplication rules for $\Sigma_1, \Sigma_2, \Sigma_3$ and Ξ_1, Ξ_2, Ξ_3. Show that each Σ matrix commutes with each Ξ matrix.

14-5. Which pairs of the following matrices commute?

$$\Sigma_1, \quad \Sigma_2\Xi_1,$$

$$\Sigma_2\Xi_2, \quad \Sigma_3\Xi_3.$$

14-6. Consider the quantity represented by the matrix

$$\Sigma_2\Xi_2 + 2\Sigma_3\Xi_3.$$

What are its possible values?

14-7. Is there a state where

$$\langle\Sigma_1\rangle = \tfrac{3}{5},$$

$$\langle\Sigma_2\rangle = \tfrac{4}{5},$$

and

$$\langle\Xi_1\rangle = 1?$$

If there is, explain how it would be produced. If there is not, explain why not.

14-8. Is there a state where

$$\langle\Sigma_1\Xi_1\rangle = \tfrac{3}{5},$$

$$\langle\Sigma_2\Xi_2\rangle = \tfrac{4}{5},$$

and

$$\langle\Xi_1\rangle = 1?$$

If there is, explain how it would be produced. If there is not, explain why not.

14-9. The matrix

$$A = \begin{pmatrix} 0 & 0 & 0 & 4 \\ 0 & 3 & 0 & 0 \\ 0 & 0 & -3 & 0 \\ 4 & 0 & 0 & 0 \end{pmatrix}$$

represents a quantity whose possible values are $-4, -3, 3, 4$. Let

$$B = \begin{pmatrix} 0 & 0 & 0 & 6 \\ 0 & 8 & 0 & 0 \\ 0 & 0 & -8 & 0 \\ 6 & 0 & 0 & 0 \end{pmatrix}.$$

Calculate the matrix products AB and BA. What are the possible values of the quantity represented by B? Explain how you know that.

15 EINSTEIN'S INSTINCTS

Interest in experiments that measure two magnetic moments grew out of Albert Einstein's criticism of quantum mechanics. This is an important part of the story of Einstein's life [1].

In 1905, when he was 26, Einstein published his first great work. One paper contained the theory of the photoelectric effect. It developed the idea that light consists of quanta, or photons. This was a key contribution to the beginning of quantum physics. Next, he published his theory of Brownian motion. It was an important contribution to statistical physics and helped convince the last skeptics that atoms and molecules really exist. A third paper was the special theory of relativity. In a fourth, he derived $E = mc^2$. During that year Einstein had a full-time job at the patent office in Bern. He was married, had a small child, and did physics in his spare time. He had been unable to get an academic job and was quite isolated from the world of physics. Yet never in the history of physics has anyone been more aware of the central problems of the day and what to do about them.

It is ironic that later, when he was famous and welcomed everywhere, he became more and more separated from the mainstream of physics, an outsider and a critic. That was largely due to his attitude about quantum mechanics.

Einstein made many important contributions to the development of quantum theory. When the result appeared, in 1925 and 1926, in the form of quantum mechanics as we know it now, he was not satisfied. At first he shared in the excitement. He wrote to Mrs. Born, 7 March 1926, "The Heisenberg–Born concepts leave us all breathless, and have made a deep impression..." [2].

Soon he began to question. He wrote to Born, 4 December 1926, "Quantum mechanics is certainly imposing. But an inner voice tells me that it is not yet the real thing. The theory says a lot, but does not really bring us any closer to the secret of the 'old one.' I, at any rate, am convinced that He is not playing at dice" [2].

Einstein was not simply opposed to the use of probability. He was one of the great masters of statistical physics, especially the statistical physics of light. Then what was his objection? It was not easy for Einstein to explain, but quantum mechanics did not satisfy his instincts for what the theory should be. In 1927 he began analyzing various aspects of quantum mechanics to find something specific that could be shown wrong. He proposed discussions of imaginary experiments that defined situations where he argued the theory would fail. Niels Bohr defended quantum mechanics against Einstein's attack. Their debate continued for a decade and more [3].

The argument Einstein made most persistently, as long as he lived, was that quantum mechanics is not consistent with principles of objectivity and causality that he held to be necessary for anything to make sense. In 1947 he wrote to Born, "I cannot seriously believe in it because the theory cannot be reconciled with the idea that physics should represent a reality in time and space, free from spooky actions at a distance" [2]. This objection was made precise in a paper Einstein published with Boris Podolsky and Nathan Rosen in 1935 [4]. They considered measurements of position and momentum for two particles. The same argument was expressed in terms of spins by David Bohm in 1951 [5]. It is usually discussed in those terms now.

Here is one way to make the argument. Consider a pair of protons in a state where the total spin is zero. Suppose a measurement determines a definite value for the projection of one magnetic moment in a certain direction. Then the projection of the other magnetic moment in the same direction has the opposite value. It has a definite value. Suppose, as an alternative, a different measurement is made and a definite value is found for the projection of the first magnetic moment in a different direction. Then the projection of the other magnetic moment in that direction has a definite value. For both magnetic moments the state is different than it would be if the first measurement were made. If the projection of a magnetic moment in one direction has a definite value, the projection in a different direction does not. What is measured for one magnetic moment changes the state for the other. But the choice of what is measured can be made after the particles have separated, so there is no way that something done to one can have an effect on the other. Basic principles of causality seem to be violated.

This argument that quantum mechanics violates simple principles of causality was made tighter, although more complicated, in a paper published by Henry Stapp in 1977 [6]. His version involves measurements of the two magnetic moments for the direction combinations $ab, a\beta, \alpha b, \alpha\beta$ that we considered in the last chapter. Now we use the table of probabilities we calculated there and follow Stapp's argument. Suppose reactions produce a lot of pairs, one after the other, each in a state where the total spin is zero, with one particle going west and the other east. We number the pairs

$1, 2, 3 \ldots$ in the order they are observed. The projection of the magnetic moment that is measured for the particle that goes west is either always in the a_1, a_2, a_3 direction or always in the $\alpha_1, \alpha_2, \alpha_3$ direction, and for the particle that goes east the projection of the magnetic moment that is measured is either always in the b_1, b_2, b_3 direction or always in the $\beta_1, \beta_2, \beta_3$ direction. Any direction combination may be chosen, but once it is, that same direction combination is used for every pair. For each pair of particles, the measurements determine one of the four possible pairs of values

1, 1	1, −1
−1, 1	−1, −1

for the magnetic moments divided by μ. The first value is for the particle that went west, and the second value is for the particle that went east. The result is recorded by putting that pair's number in the box for the pair of values that was found. Thus, after measurements have been made on nine pairs, the record might look like this.

2 5	4
1 3 8	6 7 9

Suppose this is done for many many more pairs.

Suppose all the pairs are produced and separate before any measurement is made. To allow time for this, the experiment might be done out in space, so the particles could be separated a long distance, without being disturbed, before the magnetic moments are measured.

Suppose first the direction combination *ab* is chosen. We can imagine the record of the results. Each pair is represented by its number in one of the four boxes.

Suppose just before the measurements begin, after all the particles are separated, it is decided to measure the projection of the magnetic moment in the $\beta_1, \beta_2, \beta_3$ direction, instead of the b_1, b_2, b_3 direction, for the particles that go east. How does this change the record of results? We expect to get different values for the particles that go east, because a different quantity is being measured there. We assume the values are not changed for the particles that go west, because no change is made there, and these particles are already on their way west before the change is made in the east. Then the record will be changed only by pairs changing between boxes with different east values and the same west value. They can change only horizontally between boxes. For example, some pairs that would have the values $1, -1$ for the *ab* direction combination may have the values $1, 1$ for the $a\beta$ direction combination. This change in the record is indicated by the solid horizontal arrow between those boxes in the table of probabilities.

Suppose right after that change is made, before any measurements are made, it is decided to measure the projection of the magnetic moment in the $\alpha_1, \alpha_2, \alpha_3$ direction, instead of the a_1, a_2, a_3 direction, for the particles that go west. What further changes in the record of results should we expect from this? Now we expect to get different values for the particles that go west, because a different quantity is being measured there. We assume the values are not changed for the particles that go east, because no change is made there, and these particles are already on their way east before the change is made in the west. Therefore the record will be changed only by pairs changing between boxes with different west values and the same east value. They can change only vertically between boxes. For example, some pairs that would have the values $1, 1$ for the $a\beta$ direction combination may have the values $-1, 1$ for the $\alpha\beta$ direction combination. This change in the record is indicated by the vertical solid arrow between those boxes in the table of probabilities.

How many pairs have values changed all the way from $1, -1$ for the *ab* direction combination to $-1, 1$ for the $\alpha\beta$ direction combination after both direction changes are made? Not more than a fourth of all the pairs, on the average, because for these pairs, after the first change is made, the values are $1, 1$ for the $a\beta$ direction combination, and the probability for that is $\frac{1}{4}$.

Now consider what happens if the change from a_1, a_2, a_3 to $\alpha_1, \alpha_2, \alpha_3$ in the west is made first, before the change from b_1, b_2, b_3 to $\beta_1, \beta_2, \beta_3$ in the east. For the first direction change, the record of results is changed only by pairs changing vertically between boxes with different west values and the same east value. Thus some pairs that would have the values $1, -1$ for the *ab* direction combination may have the values $-1, -1$ for the αb direction combination. This change in the record is indicated by the dashed vertical arrow between those boxes. For the second direction change, the record of results is changed only by pairs changing horizontally between boxes with different east values and the same west value. Thus some pairs that would have the values $-1, -1$ for the αb direction combination may have the values $-1, 1$ for the $\alpha\beta$ direction combination. This change in the record is indicated by the horizontal dashed arrow between those boxes.

How many pairs have values changed all the way from $1, -1$ for the *ab* direction combination to $-1, 1$ for the $\alpha\beta$ direction combination after both of these direction changes are made? At least $\frac{7}{20}$ of all the pairs, on the average. We can see this in two steps. On the average, the number of pairs that have values changed from $-1, -1$ to $-1, 1$ in the direction change from αb to $\alpha\beta$, along the horizontal dashed arrow, is at least

$$\tfrac{4}{10} - \tfrac{1}{20} = \tfrac{7}{20}$$

of all the pairs, because $\frac{4}{10}$ of the pairs have values $-1, 1$ for the $\alpha\beta$ combination and only $\frac{1}{20}$ have values $-1, 1$ for the αb combination. All the pairs that have values $-1, -1$ for the αb combination have values changed from $1, -1$ to $-1, -1$ in the direction change from ab to αb, along the vertical dashed arrow, because there are no pairs with values $-1, -1$ for the ab combination.

The two direction changes can be made independently. The particles can be separated far enough that signals traveling at the speed of light do not have time to bring information to the east about a direction change made in the west, or to the west about a change made in the east, until after all the magnetic moments are measured. It should make no difference which change is made first. We should get the same result either way. One way we found a number must be not more than $\frac{1}{4}$. The other way we found it must be at least $\frac{7}{20}$. These are inconsistent. The probabilities calculated from quantum mechanics are evidently inconsistent with the assumption that the direction change made in the west does not change the values in the east and the direction change made in the east does not change the values in the west.

To make this argument, we have to assume there is a value for the projection of each magnetic moment in each direction. The form of the argument not only shows how natural it is to assume this implicitly, but also shows how the conclusion depends on it.

REFERENCES

1. A. Pais, *Subtle Is the Lord: The Science and the Life of Albert Einstein*. Oxford, 1982.
2. *The Born–Einstein Letters*, translated by Irene Born. Walker, New York, 1971.
3. N. Bohr, "Discussion with Einstein on Epistemological Problems in Atomic Physics," in *Albert Einstein: Philosopher-Scientist*, edited by P. A. Schilpp. The Library of Living Philosophers, Evanston, Illinois, 1949, p. 199.
4. A. Einstein, B. Podolsky, and N. Rosen, *Phys. Rev.* **47**, 777 (1935).
5. D. Bohm, *Quantum Theory*. Prentice-Hall, Englewood Cliffs, New Jersey, 1951, pp. 614–623.
6. H. P. Stapp, *Nuovo Cimento* **40B**, 191 (1977).

16 BELL'S INEQUALITIES

Here is another way to look at the argument of Einstein, Podolsky, and Rosen. Consider a pair of protons that are in a state where the total spin is zero and are moving away from each other. We can measure a projection of the magnetic moment of one particle without disturbing that particle at all. We just wait until the particles are well separated and measure the projection in the same direction for the other particle. The value of the projection for the first particle must be the opposite of the value we measure for the other particle, because the total spin is zero. Thus the projection of the magnetic moment of one particle is determined by a measurement made on the other particle. We can measure the projection in any direction we choose. This is taken to mean there is a value for the projection in each direction; if we choose to measure it, we will find it, without disturbing it, so it must be there whether we choose to measure it or not. Since quantum mechanics does not allow definite values for projections of a magnetic moment in more than one direction, this argument is presented as a demonstration that quantum mechanics is incomplete.

Let us follow this line of thinking and see where it leads. Suppose there is a definite value for the projection of each magnetic moment in each direction. We had to assume this to make the argument at the end of the last chapter. Now we consider another argument that makes this assumption explicit. A particularly clear way to state it was discovered by David Mermin [1]. We follow that, using the last example worked out in Chapter 14.

As before, suppose reactions produce a lot of pairs, one after the other, each in a state where the total spin is zero, with the two particles going in opposite directions. For each particle, a measurement determines the projection of the magnetic moment in the direction of one of the three vectors $\gamma_1, \gamma_2, \gamma_3$ or $\delta_1, \delta_2, \delta_3$ or $\varepsilon_1, \varepsilon_2, \varepsilon_3$ that we considered at the end of Chapter

14. We refer to these as the γ, δ, and ε directions. Now suppose the choice of one of these three directions is made independently for each particle and for each pair and is random. Since there are three possible directions for each magnetic moment, there are nine possible direction combinations for a pair. All nine combinations are equally probable. On the average, any one will occur as often as any other.

In three of the combinations, the directions are the same for the two particles, so the projections of the two magnetic moments divided by μ should have opposite values; the pair of values should be either $1, -1$ or $-1, 1$. By observing that this happens whenever the directions are the same, the experimenter can verify that each pair is produced in a state where the total spin is zero.

In six of the combinations, the directions are different for the two particles. How often are the values opposite, either $1, -1$ or $-1, 1$, when the two directions are different? Suppose we assume there is a value for the projection of each magnetic moment in each of the three directions. For each direction, the values for the two particles are opposite, because the total spin is zero. For example, if the three projections of the magnetic moment divided by μ in the $\gamma, \delta, \varepsilon$ directions have the values $1, 1, -1$

γ	δ	ε
1	1	-1

for one particle, then they have the values $-1, -1, 1$

γ	δ	ε
-1	-1	1

for the other particle. Suppose these are the values. Then the pair of measured values will be $1, 1$ for the γ, ε and δ, ε direction combinations, $-1, -1$ for the ε, γ and ε, δ combinations, and $1, -1$ for the γ, δ and δ, γ combinations. The two values are the same for four and opposite for two of the six direction combinations where the two directions are different. Since all these direction combinations are equally probable, the two measured values should be opposite one third of the time, on the average, when the two directions are different.

There are eight sets of values

γ	δ	ε
1	1	-1
1	-1	1
-1	1	1
1	-1	-1
-1	1	-1
-1	-1	1
1	1	1
-1	-1	-1

for the three projections of a magnetic moment divided by μ in the three directions. We just considered the first of these eight sets, but we can see that the result will be the same for any of the first six sets because all that matters is that the values are identical for two directions and opposite for the third; the measured values for the two particles should be opposite one third of the time, on the average, when the two directions are different.

For either of the last two sets of values, the two measured values will be opposite for any direction combination. For example, if the three projections of the magnetic moment divided by μ in the $\gamma, \delta, \varepsilon$ directions have the values $1, 1, 1$

γ	δ	ε
1	1	1

for one particle, then they have the values $-1, -1, -1$

γ	δ	ε
-1	-1	-1

for the other particle, so the measured values for the two particles are opposite for any direction combination.

Altogether, no matter which of the eight sets of values we have for the projections in the three directions, the two measured values should be opposite, either $1, -1$ or $-1, 1$, at least one third of the time, on the average, when the two directions are different. That is the conclusion of this argument.

At the end of Chapter 14 we used quantum mechanics to calculate the probability that the two measured values are opposite. There we found that the answer is $\frac{1}{4}$. We have two inconsistent results. Quantum mechanics predicts the two measured values will be opposite one fourth of the time, on the average. The argument we just followed predicts they will be opposite at least one third of the time. One of these must be wrong.

The conclusion of the argument is one of the statements that are called Bell inequalities. The first of them were found by John S. Bell in 1964 [2]. The argument constructed by Mermin provides an example of a Bell inequality, the statement that the two measured values should be opposite at least one third of the time. Three characteristics make Bell inequalities important [3].

First, Bell inequalities are obtained from ideas about objective reality and causality that appear to be good common sense. In Mermin's example the Bell inequality is the conclusion of an argument that is based, as described at the beginning of this chapter, on two primary assumptions. One is that a measured value of a projection of a magnetic moment is part of an objective reality that exists whether we measure it or not. The other is that a measurement made on one particle does not disturb the other if the particles are separated far enough.

Second, Bell inequalities conflict with quantum mechanics. In Mermin's example the Bell inequality says the measured values should be opposite at least one third of the time and quantum mechanics says they should be opposite one fourth of the time.

Third, Bell inequalities have been tested by experiments and found to be wrong. Magnetic moments have been measured for pairs of protons, as we have described [4]. More experiments have been done with pairs of photons, measuring polarizations instead of magnetic moments [3, 5, 6]. The results agree with quantum mechanics but not with the Bell inequalities.

This is an important test of quantum mechanics in an area where there was substantial doubt. Moreover, something can be learned, without even considering quantum mechanics, from the disagreement between Bell inequalities and experiments; there must be something wrong with the assumptions that imply Bell inequalities. We have described the two primary assumptions. They are the basis for the argument of Einstein, Podolsky and Rosen. Where could they be wrong?

Is there anything wrong with the assumption that a measurement on one particle does not affect the other? If there were an effect, physicists would assume it travels at the speed of light or slower. One experiment is set up to make that impossible; for each pair, the two directions are chosen after the particles have separated and when there is not enough time for a signal traveling at the speed of light to get from one to the other before the measurements are completed [7]. Still the results agree with quantum mechanics but not with the Bell inequalities.

If there were an effect, we would expect it to be smaller and smaller when the particles are farther and farther apart. What are observed in the experiments are correlations, between magnetic moments or polarizations, that exceed limits set by the Bell inequalities. In Mermin's example, the Bell inequality says the two measured values should not be the same more than two thirds of the time, on the average, when the two directions are different, and quantum mechanics, which agrees with experiments, says they are the same three fourths of the time. Neither experiments nor theoretical arguments indicate that the correlations depend on the distance between the particles.

There are correlations that are not expected from the assumption that a measurement on one particle has no effect on the other. But that does not imply there is an effect. It does not imply there is a signal. The correlations cannot be used to send signals. To see why, suppose we try it. A device that produces pairs of particles with total spin zero is set between us so that one particle comes to me and the other to you. I can measure a projection of the magnetic moment of the particle that comes to me and can choose the direction of the projection I measure. You can do the same with the particle that comes to you. I can write out a message in a binary language and choose one direction to represent 0 and another direction to represent 1. When I measure a projection of a magnetic moment in the first direction, it represents a 0. When I measure a projection in the other direction, it represents a 1. If what I do to my particle has an effect on yours, you should be able to tell what I do to mine from measurements you make on yours. You should be able to read my message. You can measure the projection of the magnetic moment of your particle in any direction you choose. What will you find?

In Chapter 14 we showed that the mean value of the projection of either magnetic moment in any direction is zero for a state where the total spin is zero. That means there are equal probabilities $\frac{1}{2}$ for the two values μ and $-\mu$. For any direction you choose, you will find these two values occurring randomly with equal probability. That is all you will find. It will not be changed by the measurement I make. From your measurements you will not be able to learn anything about mine. No matter what directions I choose, I will not be able to send you a signal. Correlations will be evident only when we compare the record of your measurements with the record of mine.

There are correlations, but they do not represent a signal. They cannot be traced to an effect of one measurement on the other. They are created when the two particles are produced in a state where the total spin is zero and do not disappear when the particles separate. The state where the total spin is zero involves both particles. They are not independent.

To analyze the failure of Bell inequalities further, we have to consider the other assumption, that a measured value is part of an objective reality that exists whether we choose to measure it or not. That is needed for the argument that there is a value for the projection of a magnetic moment in each direction, which supports the claim that quantum mechanics is "incomplete." Between publication of the paper of Einstein, Podolsky, and Rosen in 1935 and Bell's paper in 1964, there were attempts to find a "complete" theory that would reproduce the predictions of quantum mechanics. It would include the "hidden variables" that are excluded from quantum mechanics. For example, for a magnetic moment it would allow a value for the projection in each direction. It was hoped that the theory would agree with the ideas about objective reality and causality that are now known to imply Bell inequalities. When Bell showed that the theory would imply his inequalities, he showed that this hope was futile; since Bell inequalities conflict with quantum mechanics, the theory could not reproduce the predictions of quantum mechanics. Now that experiments have shown Bell inequalities are wrong, we know the theory would also be wrong.

This forces us to change our thinking about the objectivity of reality and the independence of separated things. The old common sense is not adequate. As yet there is no new common sense. Views debated in "the reality marketplace," as Heinz Pagels describes it, still reflect "our attempt to characterize what quantum reality is not" [8]. We have quantum mechanics itself to characterize what quantum reality is. For example, in quantum mechanics there is a probability for each possible value of a projection of a magnetic moment. That is all. There is not a definite value for the projection in each direction. Philosophical interpretations will be tested, in time, by what can be made of them, whether they lead in turn to new understanding or new physics. Our current understanding developed over a period of time that extended from Einstein's initial arguments in 1927 and the paper of Einstein, Podolsky, and Rosen in 1935 to Bell's paper in 1964 and experiments done since 1971. The story may not be finished yet. Who can say what the next step may be?

In 1944, Einstein wrote to Born, "We have become antipodean in our scientific expectations. You believe in the God who plays dice, and I in complete law and order in a world which objectively exists.... No doubt the day will come when we will see whose instinctive attitude was the correct one" [9]. That day has come. Experiments have shown that Einstein's instincts were wrong.

In 1949, Born published this view of Einstein's position:

> He has seen more clearly than anyone before him the statistical background of the laws of physics, and he was a pioneer in the struggle for conquering the wilderness of quantum phenomena. Yet later, when out of his own work a synthesis of statistical and quantum principles emerged which seemed to be acceptable to almost all physicists, he kept himself aloof and sceptical. Many of us regard this as a tragedy—for him, as he gropes his way in loneliness, and for us who miss our leader and standard bearer [10].

Perhaps that is too dark. In the light of what has followed, it is certainly not tragic that Einstein was so quick to realize how deep the implications of quantum mechanics are, or that he was so persistent with questions where progress has been made only recently.

REFERENCES

1. N. D. Mermin, *J. Philos.* **78**, 397 (1981); *Am. J. Phys.* **49**, 940 (1981). This is a delightful nontechnical presentation.
2. J. S. Bell, *Physics* **1**, 195 (1964).
3. B. d'Espagnat, *Scientific American* **241**, No. 5 (November 1979), p. 158.
4. M. Lamehi-Rachti and W. Mittig, *Phys. Rev.* **14**, 2543 (1976).
5. J. F. Clauser and A. Shimony, *Rep. Prog. Phys.* **41**, 1881 (1978).
6. A. Aspect, P. Grangier, and G. Roger, *Phys. Rev. Lett.* **49**, 91 (1982).
7. A. Aspect, J. Dalibard, and G. Roger, *Phys. Rev. Lett.* **49**, 1804 (1982).
8. H. R. Pagels, *The Cosmic Code: Quantum Physics as the Language of Nature*. Bantam Books, New York, 1983, Chaps. 12 and 13.
9. *The Born–Einstein Letters*, translated by Irene Born. Walker, New York, 1971.
10. M. Born, "Einstein's Statistical Theories," in *Albert Einstein: Philosopher–Scientist*, edited by P. A. Schilpp. The Library of Living Philosophers, Evanston, Illinois, 1949, pp. 163–164.

17 HEISENBERG'S UNCERTAINTY RELATION

In taking the first steps toward quantum mechanics, Heisenberg was looking for a theory that involves only measurable quantities‡ [1, p. 60]. The result is that these quantities are represented by matrices. Heisenberg had asked the right questions: What are the measurable quantities? How do you write equations relating them? The answer requires a surprising new concept. A quantity is something more abstract than what is measured. There is a distinction between the values that are measured and the matrices that are used to write equations relating different quantities.

Heisenberg felt he was following a good example in trying to use only measurable quantities. His model was Einstein's special theory of relativity. He told Einstein that in 1926 and was "taken aback" to find that Einstein thought it was nonsense. Einstein admitted it could be useful to focus on measurable quantities, but he insisted it makes no sense to assume that what is measurable can be specified without the theory. He convinced Heisenberg to consider that, "It is the theory which decides what we can observe" [1, pp. 62–69].

Heisenberg remembered this in 1927, when he and Bohr were struggling with the physical interpretation of the new theory, pp. 76–79. He found that for position and momentum represented by matrices Q and P that satisfy the equation

$$QP - PQ = i\hbar,$$

the theory implies that

$$\sqrt{\langle (Q - \langle Q \rangle)^2 \rangle}\sqrt{\langle (P - \langle P \rangle)^2 \rangle} \geq \tfrac{1}{2}\hbar$$

‡This principle had been expressed by Pauli and Born and discussed by the Göttingen group [3].

for any state. The product of the uncertainties for the position and momentum is never smaller than $\hbar/2$.

This was the first statement of limits quantum mechanics implies on the values two quantities can have together. The simpler example of spin was considered later, because the role of spin in atomic physics was not clear at first and the Pauli matrices were not introduced until 1927, nearly two years after the matrices for position and momentum.

To prove Heisenberg's uncertainty relation, we first show that for any real quantities represented by matrices K and L, the general rules imply

$$\langle K^2 \rangle \langle L^2 \rangle \geq \left| \tfrac{1}{2} \langle KL - LK \rangle \right|^2$$

for any state. Specifically, this follows from the general rule that

$$(K + wL)(K + w^*L) \geq 0$$

for any complex number w. To see this, first suppose $\langle L^2 \rangle$ is not 0, and let

$$w = -\frac{\langle KL \rangle}{\langle L^2 \rangle}.$$

Then

$$\begin{aligned}
\langle (K + wL)(K + w^*L) \rangle &= \langle K^2 \rangle + ww^* \langle L^2 \rangle + w^* \langle KL \rangle + w \langle LK \rangle \\
&= \langle K^2 \rangle - \langle KL \rangle w^* + w^* \langle KL \rangle - \frac{\langle KL \rangle \langle LK \rangle}{\langle L^2 \rangle} \\
&= \langle K^2 \rangle - \frac{\langle KL \rangle \langle LK \rangle}{\langle L^2 \rangle}.
\end{aligned}$$

Since $\langle (K + wL)(K + w^*L) \rangle \geq 0$ for any state, it follows that

$$\langle K^2 \rangle \langle L^2 \rangle \geq \langle KL \rangle \langle LK \rangle.$$

We have shown this is true if $\langle L^2 \rangle$ is not 0, but the result is not changed if K and L are interchanged, so it must be true also if $\langle L^2 \rangle$ is 0 but $\langle K^2 \rangle$ is not 0. If $\langle L^2 \rangle$ and $\langle K^2 \rangle$ are both 0, let

$$w = -\langle KL \rangle.$$

Then

$$\langle (K + wL)(K + w^*L) \rangle = w^*\langle KL \rangle + w\langle LK \rangle$$

$$= -\langle KL \rangle^*\langle KL \rangle - \langle KL \rangle\langle LK \rangle$$

$$= -2\langle KL \rangle^*\langle KL \rangle$$

$$= -2|\langle KL \rangle|^2.$$

Remember the general rules imply

$$\langle LK \rangle = \langle KL \rangle^*.$$

It follows that

$$-2|\langle KL \rangle|^2 \geq 0,$$

which implies $\langle KL \rangle$ is 0. Thus, for any state, whether $\langle K^2 \rangle$ and $\langle L^2 \rangle$ are 0 or not, we have shown that

$$\langle K^2 \rangle \langle L^2 \rangle \geq \langle KL \rangle \langle LK \rangle.$$

We can write

$$\langle KL \rangle = \tfrac{1}{2}\langle KL + LK \rangle + i(-i)\tfrac{1}{2}\langle KL - LK \rangle.$$

The general rules imply

$$\tfrac{1}{2}\langle KL + LK \rangle \quad \text{and} \quad -i\tfrac{1}{2}\langle KL - LK \rangle$$

are real numbers for any state. Then

$$\langle KL \rangle \langle LK \rangle = \langle KL \rangle \langle KL \rangle^*$$

$$= |\langle KL \rangle|^2$$

$$= \left|\tfrac{1}{2}\langle KL + LK \rangle + i(-i)\tfrac{1}{2}\langle KL - LK \rangle\right|^2 \geq \left|\tfrac{1}{2}\langle KL - LK \rangle\right|^2$$

because for any real numbers x and y,

$$|x + iy|^2 = x^2 + y^2 \geq y^2 = |iy|^2.$$

From the two parts together, it follows that

$$\langle K^2\rangle\langle L^2\rangle \geq \left|\tfrac{1}{2}\langle KL - LK\rangle\right|^2$$

for any state.

Then for any real quantities represented by matrices A and B, the general rules imply

$$\left\langle (A - \langle A\rangle)^2\right\rangle\left\langle (B - \langle B\rangle)^2\right\rangle \geq \left|\tfrac{1}{2}\langle AB - BA\rangle\right|^2$$

for any state. To see this, let

$$K = A - \langle A\rangle$$

$$L = B - \langle B\rangle.$$

Then

$$\begin{aligned} KL - LK &= (A - \langle A\rangle)(B - \langle B\rangle) - (B - \langle B\rangle)(A - \langle A\rangle) \\ &= AB + \langle A\rangle\langle B\rangle - A\langle B\rangle - \langle A\rangle B \\ &\quad - BA - \langle B\rangle\langle A\rangle + \langle B\rangle A + B\langle A\rangle \\ &= AB - BA \end{aligned}$$

so the statement about A and B is the same as the one we proved for K and L.

For position and momentum represented by matrices Q and P such that

$$QP - PQ = i\hbar,$$

we have shown that

$$\left\langle (Q - \langle Q\rangle)^2\right\rangle\left\langle (P - \langle P\rangle)^2\right\rangle \geq \left(\tfrac{1}{2}\hbar\right)^2.$$

Taking the square root of both sides gives Heisenberg's uncertainty relation.

It is interesting to see how measurements of position and momentum are limited in particular situations. For example, suppose a particle is located by observing it in a microscope. Then the position is measured with an accuracy that depends on the wavelength of the light. The microscope cannot distinguish distances much smaller than the wavelength λ, so the uncertainty Δx in the position measured this way is approximately

$$\Delta x = \lambda.$$

On the other hand, light consists of quanta, or photons. To apply quantum mechanics to the light, we use Einstein's relation between the frequency of the light waves and the momentum of a photon. If the frequency is ν, the energy of each photon is $h\nu$. The momentum of each photon is $h\nu/c$. The velocity of the light waves is c. That is the distance the waves move in one second. The frequency ν is the number of waves that move by in one second. The length of each wave is λ. Therefore

$$c = \nu\lambda;$$

the distance the waves move in one second is equal to the number of waves that move by in one second times the length of a wave. Then the momentum of each photon is

$$h\left(\frac{\nu}{c}\right) = \frac{h}{\lambda}.$$

We see the particle in the microscope because photons collide with it. In each collision, the momentum of the particle may be changed by an amount comparable to the momentum of the photon, so the uncertainty Δp for the particles momentum, when its position is observed, is approximately

$$\Delta p = \frac{h}{\lambda}.$$

Then

$$(\Delta x)(\Delta p) \simeq \lambda\left(\frac{h}{\lambda}\right) = h.$$

The position uncertainty can be made smaller by using light with shorter wavelength, but then the momentum uncertainty is larger.

Heisenberg worked out this example himself as soon as he discovered the uncertainty relation [1, pp. 76–79]. Many other examples have been considered [2]. Einstein and others tried to imagine ways to measure position and momentum together more accurately than the uncertainty relation allows. No way has ever been found.

Heisenberg's emphasis of measurable quantities followed some unsuccessful work that, he said, "helped to convince me of one thing: that one ought to ignore the problem of electron orbits inside the atom" [1, p. 60]. He believed the orbits are not observable. The uncertainty relations confirm that. Knowledge of an orbit would imply knowledge of both position and momentum. For an electron in a hydrogen atom, the product of the position and momentum uncertainties has to be so large, compared to the product of the orbit radius and momentum, that the orbit is only roughly defined. The numbers are given in Problem 17-3.

PROBLEMS

17-1. Find

$$\left\langle (\Sigma_1 - \langle \Sigma_1 \rangle)^2 \right\rangle,$$

$$\left\langle (\Sigma_2 - \langle \Sigma_2 \rangle)^2 \right\rangle,$$

and

$$\left\langle (\Sigma_3 - \langle \Sigma_3 \rangle)^2 \right\rangle$$

for a state where the quantity represented by Σ_3 has the definite value 1.

17-2. For every case where A and B are two of the matrices $\Sigma_1, \Sigma_2, \Sigma_3$, verify that the uncertainty relation

$$\left\langle (A - \langle A \rangle)^2 \right\rangle \left\langle (B - \langle B \rangle)^2 \right\rangle \geq \left| \tfrac{1}{2} \langle AB - BA \rangle \right|^2$$

holds for the state where the quantity represented by Σ_3 has the definite value 1.

17-3. Compare the minimum value $\hbar/2$ that Heisenberg's relation allows for the product of position and momentum uncertainties with the product of typical distance and momentum dimensions for:

(a) an oxygen molecule in air at room temperature, which has a mass about 5.3×10^{-23} g, a diameter about 3×10^{-8} cm and, on the average, a velocity around 4.4×10^4 cm/s;

(b) a potassium ion, which has a mass about 6.5×10^{-23} g, that takes about 10^{-8} s to move across the membrane of a nerve cell that is about 5×10^{-7} cm thick;

(c) an electron in a hydrogen atom, which is at a distance around 5×10^{-9} cm from the nucleus and has momentum about 2×10^{-19} g cm/s; and

(d) a pea whose diameter is 1 cm and mass is 1 g that rolls across a plate at a speed of 5 cm/s.

17-4. Suppose A represents a real quantity. Then $\langle (A - \langle A \rangle)^2 \rangle$ is real and non-negative. Show this implies

$$\langle A^2 \rangle \geq \langle A \rangle^2.$$

Show that if $\langle A^2 \rangle$ is 0, then $\langle A \rangle$ is 0 and $\langle (A - \langle A \rangle)^2 \rangle$ is 0, which means the quantity represented by A has the definite value 0.

17-5. Suppose A, B, and C represent real quantities, and

$$AB - BA = iC.$$

Show the uncertainty-relation inequalities imply $\langle C \rangle$ is 0 for any state where the quantity represented by either A or B has a definite value.

17-6. Suppose K and L represent real quantities. Show

$$\langle K^2 \rangle \langle L^2 \rangle \geq \left| \tfrac{1}{2} \langle KL + LK \rangle \right|^2$$

for any state. This implies $\langle KL + LK \rangle$ is 0 if either $\langle K^2 \rangle$ or $\langle L^2 \rangle$ is 0.

REFERENCES

1. W. Heisenberg, *Physics and Beyond: Encounters and Conversations*. Harper & Row, New York, 1971, pp. 60, 62–69, 76–79.
2. W. Heisenberg, *The Physical Principles of the Quantum Theory*, translated by C. Eckart and F. C. Hoyt. Dover, New York, 1930, Chap. II, Sec. 2.
3. J. Mehra and H. Rechenberg, *The Historical Development of Quantum Theory*, Volume 2, *The Discovery of Quantum Mechanics 1925*. Springer-Verlag, New York, 1982, Chap. V, Sec. 2.

18 QUANTIZED OSCILLATOR ENERGY

An oscillator, a particle oscillating back and forth along a line, was used as a model or example for two key calculations in the development of quantum theory. The idea of a quantum of energy was introduced by Max Planck in Berlin in 1900. Planck calculated how the energy of the radiation inside an oven is distributed between different frequencies and how it depends on the temperature of the oven. This is the same for any oven in thermal equilibrium. It does not depend on what the oven is made of. Since that was established, Planck was free to use a simple model to describe emission and absorption of radiation by the walls of the oven. What he calculated for a particular model would apply quite generally to real ovens. He had no theory then to describe the atoms and molecules that actually emit and absorb radiation. He imagined the radiation is emitted and absorbed by oscillating electric charges, as in miniature radio antennas.

Planck found a formula that agrees with experiments. It contains the number h, Planck's constant, chosen to fit the experimental data. Planck also found a surprising way to explain his formula. It corresponds to the hypothesis that the amount of energy emitted or absorbed by an oscillator must be an integral multiple of $h\nu$ where ν is the frequency of oscillation, the number of oscillations per second. Planck considered this explanation "an act of desperation" and tried for years to find a more conventional one [1].

There was no reason to expect this quantization of energy. Theories then predicted that an oscillating charge could have any amount of energy and could lose or gain any amount by emitting or absorbing radiation. For the simplest kind of motion, the energy of an oscillating particle is

$$\tfrac{1}{2}mv^2 + \tfrac{1}{2}m(2\pi\nu)^2x^2,$$

where m is the particle's mass, v is its velocity, and x is its distance from the center point on the line of motion. The two parts of the energy change as the particle oscillates, but their sum remains constant. Suppose x oscillates between $-b$ and b. When x is $-b$ or b, the particle turns around. At that instant its velocity is zero. Then the energy is

$$\tfrac{1}{2}m(2\pi\nu)^2 b^2.$$

It is reasonable to expect that for a given frequency ν, the energy can have any positive value because b can have any value.

We use

$$\omega = 2\pi\nu$$

to describe the frequency. Then

$$h\nu = \frac{h}{2\pi}2\pi\nu = \hbar\omega.$$

We also use the momentum

$$p = mv.$$

In terms of these, the energy is

$$\frac{1}{2m}p^2 + \tfrac{1}{2}m\omega^2 x^2.$$

Heisenberg used an oscillator as an example to do the first calculations in quantum mechanics. He was looking for a method of calculation that would avoid difficulties with unobservable electron orbits in the Bohr model of the atom. To develop this new method, he worked on a simple example rather than the atomic problem itself [2].

As Born soon described it, Heisenberg wrote the relation between position, momentum, and energy in terms of matrices. The position and momentum are represented by matrices Q and P that satisfy the equation

$$QP - PQ = i\hbar 1.$$

The oscillator energy is represented by the matrix

$$H = \frac{1}{2m}P^2 + \tfrac{1}{2}m\omega^2 Q^2.$$

Then, as Heisenberg discovered, the oscillator energy is quantized. It can have only the values

$$(n + \tfrac{1}{2})\hbar\omega = (n + \tfrac{1}{2})h\nu \qquad \text{for} \quad n = 0, 1, 2, 3\ldots.$$

The position and momentum are not quantized. Each has a continuous range of possible values. From Heisenberg's uncertainty relation, we see that neither

$$\langle (Q - \langle Q \rangle)^2 \rangle \quad \text{nor} \quad \langle (P - \langle P \rangle)^2 \rangle$$

can ever be zero. For any state neither the position nor the momentum has a definite value. This reflects the fact that no measurement can pick one precise value out of a continuous range of possibilities. A measurement of position or momentum is never perfectly accurate. It leaves nonzero probabilities for values in some interval of the continuous range.

The squares of the position and momentum, represented by the matrices Q^2 and P^2, are not quantized. They have continuous ranges of possible values corresponding to those of position and momentum. Yet the oscillator energy, represented by the matrix H, which is a simple combination of Q^2 and P^2, is quantized. It has no continuous range of possible values, only the discrete set of possible values $(n + \frac{1}{2})\hbar\omega$. Writing the relation between position, momentum, and energy in terms of matrices gives a result that could not be expected from the old relation written in terms of position, momentum, and energy values.

There are two ways we can see the oscillator energy is quantized. First we look at the actual matrices. Later we prove it algebraically, using only equations relating physical quantities, which are written in terms of matrices, without using what the matrices actually are.

Position and momentum can be represented by different matrices in different situations. The equations relating physical quantities are always the same. When the matrices are changed, they are all changed together so the equations relating physical quantities are not changed. The Pauli matrices used to represent spin are more standard than the matrices for position and momentum; the spin matrices are hardly ever changed.

In the oscillator problem, position and momentum are represented by the matrices

$$Q = \sqrt{\frac{\hbar}{2m\omega}} \begin{pmatrix} 0 & \sqrt{1} & 0 & 0 & \cdots \\ \sqrt{1} & 0 & \sqrt{2} & 0 & \\ 0 & \sqrt{2} & 0 & \sqrt{3} & \\ 0 & 0 & \sqrt{3} & 0 & \\ \vdots & & & & \ddots \end{pmatrix}$$

and

$$P = \sqrt{\frac{\hbar m\omega}{2}} \begin{pmatrix} 0 & -i & 0 & 0 & \cdots \\ i & 0 & -i\sqrt{2} & 0 & \\ 0 & i\sqrt{2} & 0 & -i\sqrt{3} & \\ 0 & 0 & i\sqrt{3} & 0 & \\ \vdots & & & & \ddots \end{pmatrix}.$$

These are the matrices Heisenberg and Born found. They are infinite. Every quantity that has an infinite number of possible values is represented by an infinite-dimensional matrix. A quantity represented by an $n \times n$ matrix can have at most n possible values. Let

$$R = Q - i\frac{1}{m\omega}P$$

$$S = Q + i\frac{1}{m\omega}P.$$

Then

$$Q = \tfrac{1}{2}(R + S)$$

$$P = i\frac{m\omega}{2}(R - S)$$

$$R = \sqrt{\frac{2\hbar}{m\omega}} \begin{pmatrix} 0 & 0 & 0 & 0 & \cdots \\ \sqrt{1} & 0 & 0 & 0 & \\ 0 & \sqrt{2} & 0 & 0 & \\ 0 & 0 & \sqrt{3} & 0 & \\ \vdots & & & & \ddots \end{pmatrix}$$

$$S = \sqrt{\frac{2\hbar}{m\omega}} \begin{pmatrix} 0 & \sqrt{1} & 0 & 0 & \cdots \\ 0 & 0 & \sqrt{2} & 0 & \\ 0 & 0 & 0 & \sqrt{3} & \\ 0 & 0 & 0 & 0 & \\ \vdots & & & & \ddots \end{pmatrix}$$

$$SR = \frac{2\hbar}{m\omega}\begin{pmatrix} 1 & 0 & 0 & 0 & \cdots \\ 0 & 2 & 0 & 0 & \\ 0 & 0 & 3 & 0 & \\ 0 & 0 & 0 & & \\ \vdots & & & & \ddots \end{pmatrix}$$

$$RS = \frac{2\hbar}{m\omega}\begin{pmatrix} 0 & 0 & 0 & 0 & \cdots \\ 0 & 1 & 0 & 0 & \\ 0 & 0 & 2 & 0 & \\ 0 & 0 & 0 & 3 & \\ \vdots & & & & \ddots \end{pmatrix}$$

$$SR - RS = \frac{2\hbar}{m\omega}1.$$

From this we see that

$$QP - PQ = i\hbar 1$$

because

$$QP - PQ = \tfrac{1}{2}i\frac{m\omega}{2}\left[R^2 - S^2 - RS + SR - (R^2 - S^2 + RS - SR)\right]$$

$$= i\frac{m\omega}{2}[SR - RS] = i\hbar 1.$$

The matrix that represents the oscillator energy is

$$H = \frac{1}{2m}P^2 + \tfrac{1}{2}m\omega^2 Q^2$$

$$= \frac{1}{2m}\left(\frac{im\omega}{2}\right)^2(R - S)^2 + \tfrac{1}{2}m\omega^2(\tfrac{1}{2})^2(R + S)^2$$

$$= \tfrac{1}{4}m\omega^2(RS + SR).$$

Since

$$SR = RS + \frac{2\hbar}{m\omega}1$$

we have

$$H = \tfrac{1}{4}m\omega^2\left(2RS + \frac{2\hbar}{m\omega}\right)$$

$$= \hbar\omega\left(\frac{m\omega}{2\hbar}RS + \tfrac{1}{2}\right).$$

Let

$$N = \frac{m\omega}{2\hbar}RS.$$

Then

$$H = \hbar\omega\left(N + \tfrac{1}{2}\right)$$

and

$$N = \begin{pmatrix} 0 & 0 & 0 & 0 & \cdots \\ 0 & 1 & 0 & 0 & \\ 0 & 0 & 2 & 0 & \\ 0 & 0 & 0 & 3 & \\ \vdots & & & & \ddots \end{pmatrix}.$$

Consider the quantity represented by N. It is what you get if you divide the oscillator energy by $\hbar\omega$ and then subtract $\frac{1}{2}$. Knowing the value of this quantity is equivalent to knowing the value of the energy. We show the only values this quantity can have are the numbers

$$n = 0, 1, 2, 3, \ldots .$$

That means the possible values of the oscillator energy are $(n + 1/2)\hbar\omega$. We have

$$N = 0 \cdot I_0 + 1 \cdot I_1 + 2I_2 + 3I_3 + \cdots ,$$

where

$$I_0 = \begin{pmatrix} 1 & 0 & 0 & 0 & \cdots \\ 0 & 0 & 0 & 0 & \\ 0 & 0 & 0 & 0 & \\ 0 & 0 & 0 & 0 & \\ \vdots & & & & \ddots \end{pmatrix},$$

$$I_1 = \begin{pmatrix} 0 & 0 & 0 & 0 & \cdots \\ 0 & 1 & 0 & 0 & \\ 0 & 0 & 0 & 0 & \\ 0 & 0 & 0 & 0 & \\ \vdots & & & & \ddots \end{pmatrix},$$

$$I_2 = \begin{pmatrix} 0 & 0 & 0 & 0 & \cdots \\ 0 & 0 & 0 & 0 & \\ 0 & 0 & 1 & 0 & \\ 0 & 0 & 0 & 0 & \\ \vdots & & & & \ddots \end{pmatrix},$$

$$I_3 = \begin{pmatrix} 0 & 0 & 0 & 0 & \cdots \\ 0 & 0 & 0 & 0 & \\ 0 & 0 & 0 & 0 & \\ 0 & 0 & 0 & 1 & \\ \vdots & & & & \ddots \end{pmatrix},$$

and so on. The meaning of the quantities represented by the matrices $I_0, I_1, I_2, I_3, \ldots$ will be clear when we see how their values are related to the value of the quantity represented by N. The product of two different matrices $I_0, I_1, I_2, I_3, \ldots$ is zero:

$$I_0 I_1 = 0 = I_1 I_0, \qquad I_0 I_2 = 0 = I_2 I_0,$$

$$I_0 I_3 = 0 = I_3 I_0, \qquad I_1 I_2 = 0 = I_2 I_1,$$

$$I_1 I_3 = 0 = I_3 I_1, \qquad I_2 I_3 = 0 = I_3 I_2,$$

and so on. The matrices $I_0, I_1, I_2, I_3, \ldots$ commute with each other. They represent quantities that have definite values together. These values determine the value of the quantity represented by N. For each of these matrices I_j, we have

$$I_j^2 = I_j.$$

The quantity represented by I_j is the same as its square. If x is a value of this quantity, then

$$x^2 = x.$$

The only possible values are 0 and 1. Therefore

$$0 \le \langle I_j \rangle \le 1$$

for any state. We also have

$$I_0 + I_1 + I_2 + I_3 + \cdots = 1,$$

so

$$\langle I_0 \rangle + \langle I_1 \rangle + \langle I_2 \rangle + \langle I_3 \rangle + \cdots = 1$$

for any state. From this we see that if one of $\langle I_0 \rangle, \langle I_1 \rangle, \langle I_2 \rangle, \langle I_3 \rangle, \ldots$ is 1, then the others are 0. If one of the quantities represented by $I_0, I_1, I_2, I_3, \ldots$ has the value 1, then all the others have the value 0. Then the quantity represented by N has a definite value. For example, if the quantity represented by I_3 has the value 1, then the quantities represented by $I_0, I_1, I_2, I_4, \ldots$ all have the value 0, and the quantity represented by N has the value 3. In general, if the quantity represented by I_n has the value 1, where n is one of the numbers $0, 1, 2, 3, \ldots,$ then all the quantities represented by the matrices I_j for $j \neq n$ have the value 0, and the quantity represented by N has the value n.

Now we show these are the only definite values the quantity represented by N can have. For any state, consider the matrix

$$N - \langle N \rangle = \begin{pmatrix} 0 - \langle N \rangle & 0 & 0 & 0 & \cdots \\ 0 & 1 - \langle N \rangle & 0 & 0 & \\ 0 & 0 & 2 - \langle N \rangle & 0 & \\ 0 & 0 & 0 & 3 - \langle N \rangle & \\ \vdots & & & & \ddots \end{pmatrix}.$$

We have

$$N - \langle N \rangle = (0 - \langle N \rangle)I_0 + (1 - \langle N \rangle)I_1 + (2 - \langle N \rangle)I_2 + (3 - \langle N \rangle)I_3 + \cdots .$$

The square of this matrix is

$$(N - \langle N \rangle)^2 = (0 - \langle N \rangle)^2 I_0 + (1 - \langle N \rangle)^2 I_1 + (2 - \langle N \rangle)^2 I_2 + (3 - \langle N \rangle)^2 I_3 + \cdots .$$

Therefore, for any state,

$$\left\langle (N - \langle N \rangle)^2 \right\rangle = (0 - \langle N \rangle)^2 \langle I_0 \rangle + (1 - \langle N \rangle)^2 \langle I_1 \rangle + (2 - \langle N \rangle)^2 \langle I_2 \rangle + (3 - \langle N \rangle)^2 \langle I_3 \rangle + \cdots .$$

If the quantity represented by N has a definite value, then $\langle (N - \langle N \rangle)^2 \rangle$ is 0.

The only way $\langle (N - \langle N \rangle)^2 \rangle$ can be 0 is if

$$\langle N \rangle = n,$$

where n is one of the numbers $0, 1, 2, 3, \ldots$ and

$$\langle I_j \rangle = 0 \qquad \text{for} \quad j \neq n.$$

Then

$$\langle I_n \rangle = 1,$$

because

$$\langle I_0 \rangle + \langle I_1 \rangle + \langle I_2 \rangle + \langle I_3 \rangle + \cdots = 1,$$

so the quantity represented by I_n has the value 1, and the quantity represented by N has the value n.

The quantity represented by I_n has the value 1 if the quantity represented by N has the value n. It has the value 0 if the quantity represented by N has any other definite value.

We showed the only definite values the quantity represented by N can have are

$$n = 0, 1, 2, 3, \ldots .$$

This means the only definite values the oscillator energy can have are

$$(n + \tfrac{1}{2})\hbar\omega.$$

In fact, these are the only possible values. There is no continuous range of possible values. We can extend our argument to show that. Suppose there were a continuous range of possible values. Then the quantity represented by N would have a possible value that is not one of the numbers $0, 1, 2, 3, \ldots$. Let n be the number $0, 1, 2, 3, \ldots$ that is closest to that value. There would be states for which $\langle N \rangle$ is that value. The uncertainty would not be zero for any of these states, but for some the uncertainty would be less than half the difference between n and $\langle N \rangle$, so that

$$\left\langle (N - \langle N \rangle)^2 \right\rangle \leq \left[\tfrac{1}{2}(n - \langle N \rangle)\right]^2.$$

The value could be measured with that accuracy. But our formula shows that

$$\left\langle (N - \langle N \rangle)^2 \right\rangle \geq (n - \langle N \rangle)^2 (\langle I_0 \rangle + \langle I_1 \rangle + \langle I_2 \rangle + \langle I_3 \rangle \cdots)$$

or

$$\left\langle (N - \langle N \rangle)^2 \right\rangle \geq (n - \langle N \rangle)^2$$

for every state where $\langle N \rangle$ is the value we are considering, because none of the numbers

$$(0 - \langle N \rangle)^2, \quad (1 - \langle N \rangle)^2, \quad (2 - \langle N \rangle)^2, \quad (3 - \langle N \rangle)^2, \ldots$$

is smaller than $(n - \langle N \rangle)^2$. This contradiction shows there is no continuous range of possible values. That will also be shown very simply in the algebraic argument we do next.

You may object that the matrices Q and P were chosen so the matrix representing the oscillator energy turns out to be $\hbar\omega(N + \frac{1}{2})$, where N is the simple matrix that gives the answer. That is true. However, we also showed these matrices Q and P satisfy the equation

$$QP - PQ = i\hbar.$$

From that we obtain Heisenberg's uncertainty relation, which implies neither position nor momentum is quantized.

Now we use an algebraic method to show the oscillator energy is quantized. We use equations relating different matrices without using what the matrices actually are. We assume

$$QP - PQ = i\hbar.$$

The matrices R and S are defined in terms of Q and P the same as before. Then

$$\begin{aligned} RS - SR &= \frac{i}{m\omega}2(QP - PQ) \\ &= -\frac{2\hbar}{m\omega}, \end{aligned}$$

so

$$RS = SR - \frac{2\hbar}{m\omega},$$

and, as before, the matrix representing the oscillator energy is

$$H = \tfrac{1}{2}m\omega^2 RS + \tfrac{1}{2}\hbar\omega.$$

It follows that

$$\begin{aligned} RSS &= SRS - \frac{2\hbar}{m\omega}S \\ \tfrac{1}{2}m\omega^2 RSS &= S\tfrac{1}{2}m\omega^2 RS - \hbar\omega S \\ HS &= SH - \hbar\omega S \\ HS &= S(H - \hbar\omega) \\ RS &= \frac{2}{m\omega^2}(H - \tfrac{1}{2}\hbar\omega). \end{aligned}$$

By replacing HS with $S(H - \hbar\omega)$ repeatedly, we get

$$R^{n+1}S^{n+1} = R^n RSS^n$$

$$= R^n \frac{2}{m\omega^2}\left(H - \tfrac{1}{2}\hbar\omega\right)S^n$$

$$= R^n S^n \frac{2}{m\omega^2}\left(H - n\hbar\omega - \tfrac{1}{2}\hbar\omega\right)$$

for $n = 0, 1, 2, 3, \ldots$. For example,

$$R^3S^3 = R^2 \frac{2}{m\omega^2}\left(H - \tfrac{1}{2}\hbar\omega\right)SS$$

$$= R^2 S \frac{2}{m\omega^2}\left(H - \hbar\omega - \tfrac{1}{2}\hbar\omega\right)S$$

$$= R^2 SS \frac{2}{m\omega^2}\left(H - \hbar\omega - \hbar\omega - \tfrac{1}{2}\hbar\omega\right)$$

$$= R^2 S^2 \frac{2}{m\omega^2}\left(H - 2\hbar\omega - \tfrac{1}{2}\hbar\omega\right).$$

For $n = 0$, we mean

$$R^0 = 1,$$

$$S^0 = 1.$$

Starting with $n = 0$, we can work up step by step to get R^nS^n in terms of H for any n. For example,

$$R^3S^3 = R^2S^2 \frac{2}{m\omega^2}\left(H - 2\hbar\omega - \tfrac{1}{2}\hbar\omega\right)$$

$$= RS \frac{2}{m\omega^2}\left(H - \hbar\omega - \tfrac{1}{2}\hbar\omega\right) \frac{2}{m\omega^2}\left(H - 2\hbar\omega - \tfrac{1}{2}\hbar\omega\right)$$

$$= \frac{2}{m\omega^2}\left(H - \tfrac{1}{2}\hbar\omega\right) \frac{2}{m\omega^2}\left(H - \hbar\omega - \tfrac{1}{2}\hbar\omega\right) \frac{2}{m\omega^2}\left(H - 2\hbar\omega - \tfrac{1}{2}\hbar\omega\right).$$

Thus we see R^nS^n represents a quantity obtained from the oscillator energy.

For each value of the oscillator energy there is a corresponding value for the quantity represented by R^nS^n.

The general rules imply R^nS^n represents a non-negative real quantity. From the rule that

$$(B + iD)(B - iD)$$

represents a non-negative real quantity if B and D represent real quantities, it follows that RS represents a non-negative real quantity. From the rule that

$$(K + iL)G(K - iL)$$

represents a non-negative real quantity if K and L represent real quantities and G represents a non-negative real quantity, it follows that

$$R^{n+1}S^{n+1} = RR^nS^nS$$

represents a non-negative real quantity if R^nS^n does. From these, working up step by step, we can see R^nS^n represents a non-negative real quantity for any n.

Now we can determine the values the oscillator energy can have. Suppose the value is ε for some state. Then the value of the quantity represented by

$$H - n\hbar\omega - \tfrac{1}{2}\hbar\omega$$

is

$$\varepsilon - n\hbar\omega - \tfrac{1}{2}\hbar\omega.$$

This eventually becomes negative as n goes up. A negative value for this quantity and a positive value for the quantity represented by R^nS^n imply a negative value for the quantity represented by $R^{n+1}S^{n+1}$. But the quantity represented by $R^{n+1}S^{n+1}$ cannot have a negative value, and the quantity represented by R^nS^n does have a positive value when n is 0. The only possibility is that, for some n, the value of the quantity represented by $R^{n+1}S^{n+1}$ is 0; then the quantities represented by $R^{n+2}S^{n+2}$, $R^{n+3}S^{n+3}$, $R^{n+4}S^{n+4}, \ldots$ all have the value 0. There must be some n, which is one of the numbers $0, 1, 2, 3, \ldots$, where this happens; the value of the quantity represented by R^nS^n is not 0, but the value of the quantity represented by

$R^{n+1}S^{n+1}$ is 0. That implies the value of the quantity represented by

$$H - n\hbar\omega - \tfrac{1}{2}\hbar\omega$$

is 0, so

$$\varepsilon = n\hbar\omega + \tfrac{1}{2}\hbar\omega$$

$$= \left(n + \tfrac{1}{2}\right)\hbar\omega.$$

These are the only values the oscillator energy can have.

There is no continuous range of possible values. The value of the quantity represented by $H - n\hbar\omega - \frac{1}{2}\hbar\omega$ must be exactly zero.

Dirac used algebraic methods from the beginning. He explains:

> I saw that the noncommutation was really the dominant characteristic of Heisenberg's new theory. It was really more important than Heisenberg's idea of building up the theory in terms of quantities closely connected with experimental results. So I was led to concentrate on the idea of noncommutation and to see how the ordinary dynamics which people had been using until then should be modified to include it [3].

Dirac invented the word *commute* [4]. He recalls further:

> I was dealing with these new variables, the quantum variables, and they seemed to me to be some very mysterious physical quantities, and I invented a new word to describe them. I called them q-numbers, and the ordinary variables of mathematics I called c-numbers, to distinguish them. The letter q stands for quantum,... and the letter c stands for classical or maybe commuting. Then I proceeded to build up a theory of these q-numbers....
>
> Now, I did not know anything about the real nature of q-numbers. Heisenberg's matrices I thought were just an example of q-numbers; maybe q-numbers were really something more general. All that one knew about q-numbers was that they obeyed an algebra satisfying the ordinary axioms except for the commutative axiom of multiplication.... I did not bother at all about finding a precise mathematical nature for q-numbers... [4].

In fact, all physical quantities in quantum mechanics are represented by matrices, so all quantum variables are matrices. We shall continue to call them matrices, even when we do calculations algebraically and use only equations that relate different matrices without using what the matrices actually are.

PROBLEMS

18-1. Suppose in a system described by 4×4 matrices there is a real quantity represented by the matrix

$$\begin{pmatrix} 3 & 0 & 0 & 0 \\ 0 & 4 & 0 & 0 \\ 0 & 0 & 6 & 0 \\ 0 & 0 & 0 & 8 \end{pmatrix}.$$

What values can this quantity have? How do you know?

18-2. Remember

$$\left\langle (Q - \langle Q \rangle)^2 \right\rangle = \langle Q^2 \rangle - \langle Q \rangle^2$$

$$\left\langle (P - \langle P \rangle)^2 \right\rangle = \langle P^2 \rangle - \langle P \rangle^2.$$

Use this to show Heisenberg's uncertainty relation implies

$$\langle Q^2 \rangle \langle P^2 \rangle \geq \left(\frac{\hbar}{2} \right)^2$$

for any state. By working out the square in

$$\left(\langle Q^2 \rangle - \frac{\hbar}{2m\omega} \right)^2 \geq 0$$

show that

$$\langle Q^2 \rangle \left(m\hbar\omega - m^2\omega^2 \langle Q^2 \rangle \right) \leq \left(\frac{\hbar}{2} \right)^2$$

for any state and show the equality holds in the latter only if it holds in the former. Now consider a state where the oscillator energy has the value $\hbar\omega/2$. Show that

$$\langle Q^2 \rangle \langle P^2 \rangle = \langle Q^2 \rangle \left(m\hbar\omega - m^2\omega^2 \langle Q^2 \rangle \right)$$

for that state. Use all this to find

$$\langle Q^2 \rangle, \qquad \langle P^2 \rangle,$$

$$\tfrac{1}{2}m\omega^2\langle Q^2 \rangle, \qquad \frac{1}{2m}\langle P^2 \rangle,$$

$$\langle Q \rangle, \qquad \langle P \rangle$$

for that state.

18-3. Show that

$$H = \tfrac{1}{2}m\omega^2 SR - \tfrac{1}{2}\hbar\omega.$$

Use this to show that

$$RH - HR = -\hbar\omega R.$$

Show also that

$$HS - SH = -\hbar\omega S.$$

Use the last two equations and the result of Problem 17-5 to find $\langle R \rangle$, $\langle S \rangle$ and $\langle Q \rangle$, $\langle P \rangle$ for any state where the oscillator energy has a definite value.

18-4. Show that

$$R^2H - HR^2 = -2\hbar\omega R^2$$

and

$$HS^2 - S^2H = -2\hbar\omega S^2.$$

Use these and the result of Problem 17-5 to find $\langle R^2 \rangle$, $\langle S^2 \rangle$, $\langle Q^2 \rangle$, $\langle P^2 \rangle$ and $(\tfrac{1}{2})m\omega^2\langle Q^2 \rangle$, $(1/2m)\langle P^2 \rangle$ for the state where the oscillator energy has the value $(n + \tfrac{1}{2})\hbar\omega$.

REFERENCES

1. E. Segrè, *From X-Rays to Quarks: Modern Physicists and Their Discoveries*. W. H. Freeman, San Francisco, 1980, Chap. 4.
2. W. Heisenberg, *Physics and Beyond: Encounters and Conversations*. Harper & Row, New York, 1971, pp. 60–61.
3. P. A. M. Dirac, *The Development of Quantum Theory*. Gordon and Breach, New York, 1971, p. 23.
4. P. A. M. Dirac, in *History of Twentieth Century Physics*, edited by C. Weiner. Academic Press, New York, 1977, p. 129.

19 BOHR'S MODEL

The central problems the creators of quantum mechanics hoped to solve were brought into focus by the Bohr model of the atom. It is the theory physicists used to understand and relate experimental facts about atoms before quantum mechanics was developed. One important fact is that the light atoms emit has only certain frequencies. Another is that atoms are stable.

To explain experiments involving collisions of ions with atoms, Ernest Rutherford suggested in 1911 that an atom consists of a nucleus surrounded by electrons moving in orbits like planets [1]. Theories then predicted an electron orbit would continually change as the electron emits radiation and loses energy. The stability of the atom means this does not happen. It also means the orbit is not changed in most collisions of the atom with other atoms and molecules.

According to the model proposed by Niels Bohr in 1913, the energy in an atom can have only certain values. When the atom emits radiation, the energy changes from an initial value ε_i to a final value ε_f such that

$$\varepsilon_i - \varepsilon_f = h\nu,$$

where ν is the frequency of the radiation. Thus Bohr combined Rutherford's picture of the atom with the idea of Planck and Einstein that radiation of frequency ν is emitted in quanta of energy $h\nu$. Emission of radiation is not viewed as a continuous process, but as a sudden event like a radioactive decay. It has to be described statistically. We cannot predict exactly when it will happen. We can only calculate the probability that it will happen in a given time.

When the energy has the lowest possible value, it cannot change to a lower value, so no radiation can be emitted. Then the atom is stable. Even a collision with another atom can change the energy only if it changes it enough to reach the next possible value.

When the energy has a higher value, it can change to a lower value. Then the atom emits radiation. This was confirmed in experiments by James Franck and Gustav Hertz reported in 1914 and interpreted by Bohr in 1915. Franck and Hertz raised the energy in atoms of mercury vapor by having accelerated electrons collide with them. The electron energy was varied by changing the acceleration voltage. When the electron energy was below a certain value, the electrons lost no energy in collisions with atoms and the atoms emitted no radiation. As soon as the electron energy was above that value, the electrons did lose energy in collisions with atoms and the atoms did emit radiation. That value is evidently the difference between the lowest and the next possible value of the energy in a mercury atom. The energy in the atom was raised only when the colliding electron could supply enough energy to reach the next possible value. Then the atom emitted radiation as its energy changed back to the lower value. The energy lost by the colliding electron was found to be the same as the frequency of the radiation multiplied by h [2, 3].

The differences between the possible values of the energy in an atom determine the frequencies of radiation the atom can emit. The same idea applies to radiation emitted by an oscillator. Since an oscillator is a simpler example, we consider that first. For an oscillator the possible energy values are

$$\left(n + \tfrac{1}{2}\right) h\nu_0,$$

where ν_0 is the frequency of oscillation and n is a non-negative integer. All the energy differences are integral multiples of $h\nu_0$. When the oscillator emits a quantum of radiation of frequency ν, its energy must change by an amount

$$h\nu = \varepsilon_i - \varepsilon_f,$$

which is an integral multiple of $h\nu_0$. Therefore the frequency ν of the radiation must be an integral multiple of the frequency of oscillation ν_0. Actually, it is most likely that the energy changes only to the next possible value, so the frequency of most of the radiation is just ν_0. A molecule that consists of two atoms can behave as an oscillator with the atoms moving towards and away from each other. The observed frequencies of the radiation these molecules emit show that in many cases the energy of this oscillating motion has values that are approximately equally spaced as the simple oscillator model predicts. This was one of the first applications of quantum mechanics [4].

For an atom that has one electron, for example a hydrogen atom or an ionized helium atom, both the Bohr model and quantum mechanics predict the possible energy values are approximately

$$\varepsilon_n = -\frac{m(Ze^2)^2}{2\hbar^2 n^2}$$

for $n = 1, 2, 3, 4, \ldots$, where Ze is the electric charge of the nucleus, $-e$ is the electron charge, and

$$m = \frac{m_n m_e}{m_n + m_e}$$

where m_n is the mass of the nucleus and m_e is the electron mass. When the atom emits a quantum of radiation of frequency ν, its energy changes by an amount

$$h\nu = \varepsilon_i - \varepsilon_f.$$

Each initial energy ε_i and final energy ε_f must be one of the possible energies ε_n. That determines the frequencies ν of the radiation the atom can emit. The differences between the predicted energy values ε_n for different values of n do correspond quite accurately to the observed frequencies ν of the radiation these atoms emit. This was the first big success of the Bohr model.

In the Bohr model, equations relating different quantities are still written in terms of their values, not in terms of matrices. The orbit of the relative motion of the electron and nucleus is the result of the Coulomb force between these charges. The energy is

$$\varepsilon = \tfrac{1}{2}mv^2 - \frac{Ze^2}{r},$$

where r is the distance between the electron and the nucleus and v is their relative velocity. The energy remains constant as the particles move. It includes the energy of motion in the atom for both the electron and the nucleus. It does not include the energy of motion of the atom as a whole. We may suppose the atom is not moving. Then the momentum of the atom is zero. The momenta of the electron and nucleus are equal and opposite.

That means

$$m_e v_e = m_n v_n$$

in terms of the magnitudes v_e and v_n of the velocities of the electron and nucleus. It also means the electron and nucleus are moving in opposite directions. Their relative velocity is

$$v = v_e + v_n.$$

Then

$$v_e = \frac{m_n}{m_n + m_e} v,$$

$$v_n = \frac{m_e}{m_n + m_e} v,$$

because this is the only way to split v into two parts that satisfy the momentum equation. It follows that

$$\begin{aligned}
\tfrac{1}{2}m_e(v_e)^2 + \tfrac{1}{2}m_n(v_n)^2 &= \tfrac{1}{2}m_e\left(\frac{m_n}{m_n + m_e}\right)^2 v^2 + \tfrac{1}{2}m_n\left(\frac{m_e}{m_n + m_e}\right)^2 v^2 \\
&= \frac{1}{2}\frac{m_e m_n^2 + m_n m_e^2}{(m_n + m_e)^2} v^2 \\
&= \frac{1}{2}\frac{m_n m_e (m_n + m_e)}{(m_n + m_e)^2} v^2 \\
&= \frac{1}{2}\frac{m_n m_e}{m_n + m_e} v^2 \\
&= \tfrac{1}{2}mv^2.
\end{aligned}$$

The energy would be zero if the relative velocity v were zero and the distance r infinite. It would be positive if the electron and nucleus were to move away from each other to an infinite distance r with positive relative velocity v. If the energy is negative, they cannot move away to an infinite distance. That is how they are bound together in an atom. For negative energy the orbit is an ellipse.

In the Bohr model the energy values ε_n are obtained by imposing a "quantization condition" that is equivalent to assuming the orbital angular momentum is an integral multiple of $\hbar$. It is easy to see how this is done when the orbit is a circle. The force that makes the orbit a circle has to be equal to mv^2/r. Here it is the Coulomb force so

$$\frac{mv^2}{r} = \frac{Ze^2}{r^2}.$$

It follows that

$$mv^2 = \frac{Ze^2}{r}$$

and

$$\varepsilon = \frac{1}{2}\frac{Ze^2}{r} - \frac{Ze^2}{r}$$

$$= -\frac{1}{2}\frac{Ze^2}{r}.$$

The orbital angular momentum is

$$l = rmv.$$

We have

$$rmv^2 = Ze^2$$

so

$$\varepsilon = -\frac{1}{2}\frac{Ze^2}{r}\frac{m}{m}\frac{Ze^2}{rmv^2}$$

$$= -\frac{m(Ze^2)^2}{2r^2m^2v^2}$$

$$= -\frac{m(Ze^2)^2}{2l^2}.$$

In the Bohr model it is assumed the orbital angular momentum can have only the values

$$l = n\hbar,$$

where n is one of the positive integers $1, 2, 3, 4 \ldots$. Then the energy can have only the values

$$\varepsilon_n = -\frac{m(Ze^2)^2}{2\hbar^2 n^2}.$$

The atom has a magnetic moment proportional to the orbital angular momentum. If the orbital angular momentum is quantized, the magnetic moment must be quantized too. That was confirmed by the experiments of Stern and Gerlach. The Bohr model relates the values of magnetic moments measured by Stern and Gerlach to the observed frequencies of radiation the atom emits. They are both linked to the assumption that the orbital angular momentum is quantized.

PROBLEMS

19-1. Find formulas for the radius r, the velocity v, and the momentum $p = mv$ for the circular Bohr orbit for each $n = 1, 2, 3, 4, \ldots$. Compare rp with $\hbar/2$ for each n. You do not need to use numbers for that. Write your answers in terms of n, $\hbar$, m, and Ze^2 only.

19-2. Use the results of the last problem to find formulas for the period of revolution (that is the time one revolution takes) and the frequency of revolution (that is the number of revolutions per unit time) for the circular Bohr orbit for each n. Again, write the answers in terms of n, $\hbar$, m, and Ze^2 only.

19-3. Show that

$$-\left[\frac{1}{2(n+1)^2} - \frac{1}{2n^2}\right] = \frac{2n+1}{2n^2(n+1)^2}.$$

When n is large, that is approximately $1/n^3$. Use this to find a formula for the frequency of the radiation an atom emits, according to the Bohr model, when its energy changes from ε_{n+1} to ε_n, and show that when n is large, this is approximately the same as the frequency of revolution found in the last problem. This is an example of what Bohr called the correspondence principle; when n is large, the quantum properties are less important because $\hbar$ is small compared to the magnitude of distance and momentum, as the first problem shows, and then the frequency of the radiation is what classical electrodynamics would predict for radiation from an oscillating or periodic motion of a charge.

REFERENCES

1. E. Segré, *From X-Rays to Quarks*, W. H. Freeman, San Francisco, 1980, Chap. 6.
2. W. H. Cropper, *The Quantum Physicists*. Oxford University Press, New York, 1970, pp. 54–57.
3. L. Rosenfeld and E. Rüdinger, "The Decisive Years 1911–1918," in *Niels Bohr*: *His Life and Work as Seen by His Friends and Colleagues*, edited by S. Rozental. North-Holland, Amsterdam, 1967, pp. 65–66.
4. J. Mehra and H. Rechenberg, *The Historical Development of Quantum Theory*, Volume 3, *The Formulation of Matrix Mechanics and Its Modifications 1925–1926*. Springer-Verlag, New York, 1982, pp. 187–193.

20 ANGULAR MOMENTUM

Using quantum mechanics, we can show the orbital angular momentum is quantized. The first step is to write it in terms of position and momentum. The position of a particle in three-dimensional space is described by coordinates x_1, x_2, x_3 along three perpendicular axes. They are the projections of a position vector whose length is

$$r = \sqrt{x_1^2 + x_2^2 + x_3^2}\,.$$

The velocity of the particle also is a vector quantity. Let its projections in the three perpendicular reference directions be v_1, v_2, v_3. The momentum of the particle is the vector quantity whose projections are

$$p_1 = mv_1,$$

$$p_2 = mv_2,$$

$$p_3 = mv_3,$$

where m is the mass of the particle. The magnitudes of the velocity and momentum are

$$v = \sqrt{v_1^2 + v_2^2 + v_3^2}$$

and

$$p = \sqrt{p_1^2 + p_2^2 + p_3^2}$$

$$= mv.$$

The orbital angular momentum is the vector quantity whose projections are

$$l_1 = x_2 p_3 - x_3 p_2,$$

$$l_2 = x_3 p_1 - x_1 p_3,$$

$$l_3 = x_1 p_2 - x_2 p_1.$$

The angular-momentum vector is perpendicular to the position and momentum vectors. For example, suppose the particle is moving in the plane of the 1 and 2 axes. Then the position and momentum vectors are in that plane, so x_3 and p_3 are 0. This implies l_1 and l_2 are 0, so the angular momentum is along the 3 axis, perpendicular to the plane of motion and perpendicular to the position and momentum vectors.

Suppose the particle is moving in a circle and the position is measured from the center. Then the position and velocity vectors are perpendicular.

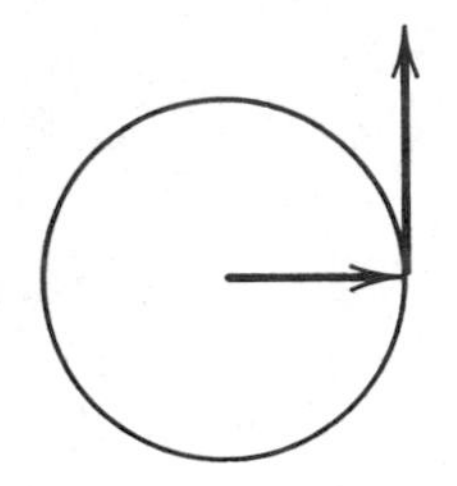

This implies the magnitude of the angular momentum is

$$rp = rmv.$$

For example, suppose the position is on the 1 axis and the momentum is in the 2 direction. Then x_2 and x_3 are 0 and p_1 and p_3 are 0, so l_1 and l_2 are 0 and

$$l_3 = x_1 p_2 = rp$$

if x_1 is positive. If x_1 were negative, or if the momentum were in the opposite direction, l_3 would be $-rp$. Whether l_3 is positive or negative depends on which way the particle goes around the circle.

In quantum mechanics the position coordinates and momentum projections are represented by matrices Q_1, Q_2, Q_3 and P_1, P_2, P_3 that satisfy the commutation relations

$$Q_1P_1 - P_1Q_1 = i\hbar,$$

$$Q_2P_2 - P_2Q_2 = i\hbar,$$

$$Q_3P_3 - P_3Q_3 = i\hbar.$$

Otherwise these matrices commute; the matrices Q_1, Q_2, Q_3 commute with each other, the matrices P_1, P_2, P_3 commute with each other, and Q_1 commutes with P_2 and P_3, and so on. There are uncertainty relations only for measurements of position and momentum in the same direction. There are no restrictions on measurements of different position coordinates together or measurements of different momentum projections together.

The orbital angular momentum is the vector quantity whose projections are represented by the matrices

$$L_1 = Q_2P_3 - Q_3P_2,$$

$$L_2 = Q_3P_1 - Q_1P_3,$$

$$L_3 = Q_1P_2 - Q_2P_1.$$

It follows that

$$L_1L_2 - L_2L_1 = i\hbar L_3,$$

$$L_2L_3 - L_3L_2 = i\hbar L_1,$$

$$L_3L_1 - L_1L_3 = i\hbar L_2.$$

For example,

$$\begin{aligned} L_1L_2 - L_2L_1 &= (Q_2P_3 - Q_3P_2)(Q_3P_1 - Q_1P_3) \\ &\quad -(Q_3P_1 - Q_1P_3)(Q_2P_3 - Q_3P_2) \\ &= Q_2P_3Q_3P_1 + Q_3P_2Q_1P_3 - Q_3P_1Q_2P_3 - Q_1P_3Q_3P_2 \\ &= Q_1P_2(Q_3P_3 - P_3Q_3) + Q_2P_1(P_3Q_3 - Q_3P_3) \\ &= (Q_1P_2 - Q_2P_1)i\hbar \\ &= i\hbar L_3. \end{aligned}$$

These equations are called angular-momentum commutation relations.

The relative position and momentum of the electron and nucleus in an atom are represented by the matrices

$$Q_j = Q_{ej} - Q_{nj}$$

and

$$\begin{aligned}
P_j &= \frac{m_n}{m_n + m_e} P_{ej} - \frac{m_e}{m_n + m_e} P_{nj} \\
&= \frac{m_n m_e}{m_n + m_e}\left(\frac{1}{m_e} P_{ej} - \frac{1}{m_n} P_{nj}\right) \\
&= m\left(V_{ej} - V_{nj}\right) \\
&= m V_j
\end{aligned}$$

for $j = 1, 2, 3$, where

$$V_{ej} = \frac{1}{m_e} P_{ej} \quad \text{and} \quad V_{nj} = \frac{1}{m_n} P_{nj}$$

represent the velocities of the electron and nucleus,

$$V_j = V_{ej} - V_{nj}$$

represent the relative velocity, and

$$m = \frac{m_n m_e}{m_n + m_e}.$$

The matrices Q_{ej}, P_{ej} for the electron commute with the matrices Q_{nj}, P_{nj} for the nucleus. Therefore

$$\begin{aligned}
Q_j P_j - P_j Q_j &= \frac{m_n}{m_n + m_e}\left(Q_{ej} P_{ej} - P_{ej} Q_{ej}\right) + \frac{m_e}{m_n + m_e}\left(Q_{nj} P_{nj} - P_{nj} Q_{nj}\right) \\
&= \left(\frac{m_n}{m_n + m_e} + \frac{m_e}{m_n + m_e}\right) i\hbar \\
&= i\hbar.
\end{aligned}$$

The relative position and momentum matrices Q_j, P_j satisfy the same commutation relations as position and momentum matrices for a particle. Therefore, the matrices L_1, L_2, L_3 made from them satisfy angular-momentum commutation relations. It is this angular momentum made from relative position and momentum that is used to calculate the energy values for the atom.

The matrices

$$S_1 = \tfrac{1}{2}\hbar\Sigma_1,$$
$$S_2 = \tfrac{1}{2}\hbar\Sigma_2,$$
$$S_3 = \tfrac{1}{2}\hbar\Sigma_3$$

that represent spin angular momentum also satisfy angular-momentum commutation relations. We have

$$\Sigma_1\Sigma_2 - \Sigma_2\Sigma_1 = i2\Sigma_3,$$
$$\Sigma_2\Sigma_3 - \Sigma_3\Sigma_2 = i2\Sigma_1,$$
$$\Sigma_3\Sigma_1 - \Sigma_1\Sigma_3 = i2\Sigma_2,$$

so

$$S_1S_2 - S_2S_1 = i\hbar S_3,$$
$$S_2S_3 - S_3S_2 = i\hbar S_1,$$
$$S_3S_1 - S_1S_3 = i\hbar S_2.$$

Angular momentum is always represented by matrices $\hbar J_1, \hbar J_2, \hbar J_3$, where J_1, J_2, J_3 satisfy commutation relations

$$J_1J_2 - J_2J_1 = iJ_3,$$
$$J_2J_3 - J_3J_2 = iJ_1,$$
$$J_3J_1 - J_1J_3 = iJ_2.$$

Later we shall see how these are related to the algebra of rotations. Now we show how these commutation relations determine the values angular momentum can have.

The matrix

$$J^2 = J_1^{\,2} + J_2^{\,2} + J_3^{\,2}$$

commutes with all the matrices J_1, J_2, J_3. For example,

$$\begin{aligned} J_3J^2 - J^2J_3 &= J_3J_1J_1 + J_3J_2J_2 - J_1J_1J_3 - J_2J_2J_3 \\ &= J_1J_3J_1 + iJ_2J_1 + J_2J_3J_2 - iJ_1J_2 \\ &\quad -J_1J_3J_1 - J_1(-iJ_2) - J_2J_3J_2 - J_2iJ_1 \\ &= 0. \end{aligned}$$

Since the matrices J_1, J_2, J_3 do not commute, projections of the angular momentum in different directions generally do not have definite values together. Since J^2 and J_3 do commute, they represent quantities that do have definite values together. We show the possible values are $j(j+1)$ for J^2 and k for J_3, where j is one of the numbers

$$0, \tfrac{1}{2}, 1, \tfrac{3}{2}, 2, \ldots,$$

and k is one of the numbers

$$-j, -j+1, -j+2, \ldots j-1, j.$$

The result would be the same if J_3 were replaced by J_1 or J_2 or the projection in any other direction.

Let

$$J_+ = J_1 + iJ_2,$$

$$J_- = J_1 - iJ_2.$$

Then

$$\begin{aligned} J_3J_+ - J_+J_3 &= iJ_2 + i(-iJ_1) \\ &= J_1 + iJ_2 \\ &= J_+ \end{aligned}$$

$$\begin{aligned} J_3J_- - J_-J_3 &= iJ_2 - i(-iJ_1) \\ &= -(J_1 - iJ_2) \\ &= -J_- \end{aligned}$$

$$\begin{aligned} J_-J_+ &= (J_1 - iJ_2)(J_1 + iJ_2) \\ &= J_1^2 + J_2^2 + i(J_1J_2 - J_2J_1) \\ &= J^2 - J_3^2 - J_3 \end{aligned}$$

$$\begin{aligned} J_+J_- &= (J_1 + iJ_2)(J_1 - iJ_2) \\ &= J_1^2 + J_2^2 - i(J_1J_2 - J_2J_1) \\ &= J^2 - J_3^2 + J_3. \end{aligned}$$

We have

$$J_3 J_+ = J_+(J_3 + 1)$$

$$J_3 J_- = J_-(J_3 - 1).$$

Using these repeatedly, we get

$$\begin{aligned}(J_-)^{n+1}(J_+)^{n+1} &= (J_-)^n J_- J_+ (J_+)^n \\ &= (J_-)^n\left[J^2 - J_3^2 - J_3\right](J_+)^n \\ &= (J_-)^n (J_+)^n\left[J^2 - (J_3 + n)^2 - (J_3 + n)\right]\end{aligned}$$

$$\begin{aligned}(J_+)^{n+1}(J_-)^{n+1} &= (J_+)^n J_+ J_- (J_-)^n \\ &= (J_+)^n\left[J^2 - J_3^2 + J_3\right](J_-)^n \\ &= (J_+)^n (J_-)^n\left[J^2 - (J_3 - n)^2 + (J_3 - n)\right]\end{aligned}$$

for $n = 0, 1, 2, 3, \ldots$. For example,

$$\begin{aligned}(J_-)^3(J_+)^3 &= (J_-)^2\left[J^2 - J_3^2 - J_3\right]J_+ J_+ \\ &= (J_-)^2 J_+\left[J^2 - (J_3 + 1)^2 - (J_3 + 1)\right]J_+ \\ &= (J_-)^2 J_+ J_+\left[J^2 - (J_3 + 1 + 1)^2 - (J_3 + 1 + 1)\right] \\ &= (J_-)^2(J_+)^2\left[J^2 - (J_3 + 2)^2 - (J_3 + 2)\right].\end{aligned}$$

Remember J_1 and J_2 commute with J^2, so J_+ and J_- commute with J^2. For $n = 0$, we mean

$$(J_+)^0 = 1$$

$$(J_-)^0 = 1.$$

Starting with $n = 0$, we can work up step by step to get $(J_-)^n(J_+)^n$ and

$(J_+)^n(J_-)^n$ in terms of J^2 and J_3 for any n. For example,

$$(J_-)^3(J_+)^3 = (J_-)^2(J_+)^2\left[J^2 - (J_3 + 2)^2 - (J_3 + 2)\right]$$

$$= J_-J_+\left[J^2 - (J_3 + 1)^2 - (J_3 + 1)\right]\left[J^2 - (J_3 + 2)^2 - (J_3 + 2)\right]$$

$$= \left[J^2 - J_3^2 - J_3\right]\left[J^2 - (J_3 + 1)^2 - (J_3 + 1)\right]\left[J^2 - (J_3 + 2)^2 - (J_3 + 2)\right].$$

Thus we see $(J_-)^n(J_+)^n$ and $(J_+)^n(J_-)^n$ represent quantities obtained from those represented by J^2 and J_3. For any possible values of the quantities represented by J^2 and J_3, there are corresponding values for the quantities represented by $(J_-)^n(J_+)^n$ and $(J_+)^n(J_-)^n$.

The general rules imply $(J_-)^n(J_+)^n$ and $(J_+)^n(J_-)^n$ represent non-negative real quantities. From the rule that

$$(B + iD)(B - iD)$$

represents a non-negative real quantity if B and D represent real quantities, it follows that J_-J_+ and J_+J_- represent non-negative real quantities. From the rule that

$$(K + iL)G(K - iL)$$

represents a non-negative real quantity if K and L represent real quantities and G represents a non-negative real quantity, it follows that

$$(J_-)^{n+1}(J_+)^{n+1} = J_-(J_-)^n(J_+)^n J_+$$

and

$$(J_+)^{n+1}(J_-)^{n+1} = J_+(J_+)^n(J_-)^n J_-$$

represent non-negative real quantities if $(J_-)^n(J_+)^n$ and $(J_+)^n(J_-)^n$ do. From these, working up step by step, we can see $(J_-)^n(J_+)^n$ and $(J_+)^n(J_-)^n$ represent non-negative real quantities for any n.

Now we can determine the values the quantities represented by J^2 and J_3 can have. Suppose their values are g and k for some state. Then the value of the quantity represented by

$$J^2 - (J_3 + n)^2 - (J_3 + n)$$

is

$$g-(k+n)^2-(k+n).$$

This eventually becomes negative as n goes up. A negative value for this quantity and a positive value for the quantity represented by $(J_-)^n(J_+)^n$ imply a negative value for the quantity represented by $(J_-)^{n+1}(J_+)^{n+1}$. But the quantity represented by $(J_-)^{n+1}(J_+)^{n+1}$ cannot have a negative value, and the quantity represented by $(J_-)^n(J_+)^n$ does have a positive value when n is 0. The only possibility is that for some n, the value of the quantity represented by $(J_-)^{n+1}(J_+)^{n+1}$ is 0; then the quantities represented by $(J_-)^{n+2}(J_+)^{n+2}$, $(J_-)^{n+3}(J_+)^{n+3}$, $(J_-)^{n+4}(J_+)^{n+4}, \ldots$ all have the value 0. There must be some n, which is one of the numbers 0, 1, 2, 3, ..., where this happens; the value of the quantity represented by $(J_-)^n(J_+)^n$ is not 0 but the value of the quantity represented by $(J_-)^{n+1}(J_+)^{n+1}$ is 0. That implies the value of the quantity represented by

$$J^2-(J_3+n)^2-(J_3+n)$$

is 0, so

$$g=(k+n)^2+(k+n).$$

Similarly, the quantity represented by

$$J^2-(J_3-n)^2+(J_3-n)$$

has the value

$$g-(k-n)^2+(k-n).$$

This also becomes negative as n goes up. Since the quantity represented by $(J_+)^{n+1}(J_-)^{n+1}$ cannot have a negative value for any n, the only possibility is that for some n, which we call n', the value of the quantity represented by

$$J^2-(J_3-n)^2+(J_3-n)$$

is 0, so

$$g=(k-n')^2-(k-n'),$$

where n' is one of the numbers 0, 1, 2, 3, ...

Combining the two formulas for g, we have

$$(k+n)^2+(k+n)=g=(k-n')^2-(k-n')$$

or

$$(k+n)(k+n+1)=g=(n'-k)(n'-k+1).$$

The quadratic equation for $n'-k$ has two solutions, either

$$n'-k=k+n$$

or

$$n'-k=-k-n-1,$$

but the second solution gives

$$n'=-n-1,$$

which implies n' is negative, so only the first solution applies. Therefore

$$g=j(j+1),$$

where

$$j=k+n$$

$$=n'-k.$$

It follows that

$$j=\tfrac{1}{2}(n'+n)$$

and

$$k=j-n$$

$$=-j+n'.$$

From this we see j must be one of the numbers

$$0,\ \tfrac{1}{2},\ 1,\ \tfrac{3}{2},\ 2,\ldots,$$

and k must be one of the numbers

$$-j,\ -j+1,\ -j+2,\ldots,\ j-1,\ j.$$

These are the only values the quantities represented by J^2 and J_3 can have. Since they can have definite values together, each value of one can occur with some value of the other. The values k are between $-j$ and j because the matrix

$$J^2 - J_3^2 = J_1^2 + J_2^2$$

represents a non-negative real quantity; if we had either

$$k \geq j + \tfrac{1}{2}$$

or

$$k \leq -j - \tfrac{1}{2},$$

then we would have

$$k^2 \geq \left(j + \tfrac{1}{2}\right)^2$$

or

$$k^2 \geq j^2 + j + \tfrac{1}{4}$$

and

$$k^2 > j(j+1),$$

which implies the quantity represented by $J^2 - J_3^2$ has a negative value.

There is no continuous range of possible values for either of the quantities represented by J^2 and J_3. The values of the quantities represented by

$$J^2 - (J_3 + n)^2 - (J_3 + n)$$

and

$$J^2 - (J_3 - n')^2 + (J_3 - n')$$

must be exactly zero.

For the orbital angular momentum, j and k are always integers, never half of odd integers. The reason for that is not evident here.

For the spin angular momentum represented by the Pauli matrices multiplied by $\hbar/2$, we have

$$J^2 = \left(\tfrac{1}{2}\Sigma_1\right)^2 + \left(\tfrac{1}{2}\Sigma_2\right)^2 + \left(\tfrac{1}{2}\Sigma_3\right)^2$$
$$= \tfrac{3}{4},$$

which is $j(j+1)$ for $j = \frac{1}{2}$. The possible values of the quantity represented by

$$J_3 = \tfrac{1}{2}\Sigma_3$$

are

$$k = -\tfrac{1}{2}, \tfrac{1}{2}.$$

PROBLEMS

20-1. Use the result of Problem 17-5 to show the angular-momentum commutation relations imply $\langle L_1 \rangle$ and $\langle L_2 \rangle$ are zero for any state where the quantity represented by L_3 has a definite value.

20-2. Work out

$$Q_1 L_3 - L_3 Q_1,$$

$$Q_2 L_3 - L_3 Q_2,$$

$$Q_3 L_3 - L_3 Q_3,$$

and write the answers in terms of Q_1, Q_2, Q_3.

Work out

$$P_1 L_3 - L_3 P_1,$$

$$P_2 L_3 - L_3 P_2,$$

$$P_3 L_3 - L_3 P_3,$$

and write the answers in terms of P_1, P_2, P_3.

What position coordinates and momentum projections do you expect can be measured together with the projection of the orbital angular momentum in the 3 direction without restriction of accuracy?

20-3. Use the result of the last problem to calculate

$$L_3\left(Q_1^2 + Q_2^2 + Q_3^2\right) - \left(Q_1^2 + Q_2^2 + Q_3^2\right)L_3$$

and

$$L_3(P_1^2 + P_2^2 + P_3^2) - (P_1^2 + P_2^2 + P_3^2)L_3.$$

Both of these can be done the same way $J_3 J^2 - J^2 J_3$ was.

20-4. Use the result of Problems 20-2 and 17-5 to show $\langle Q_1 \rangle$, $\langle Q_2 \rangle$ and $\langle P_1 \rangle$, $\langle P_2 \rangle$ are zero for any state where the quantity represented by L_3 has a definite value.

20-5. For the case where

$$J_1 = \tfrac{1}{2}\Sigma_1, \qquad J_2 = \tfrac{1}{2}\Sigma_2, \qquad J_3 = \tfrac{1}{2}\Sigma_3,$$

write out explicitly the 2×2 matrices for

$$J_+, \qquad J_-, \qquad J_-J_+, \qquad J_+J_-,$$

$$J^2 - J_3^2 - J_3, \qquad J^2 - J_3^2 + J_3$$

and

$$(J_-)^n(J_+)^n, \qquad (J_+)^n(J_-)^n$$

for $n = 2, 3, 4, \ldots$. Check that your answers give

$$J_-J_+ = J^2 - J_3^2 - J_3$$

and

$$J_+J_- = J^2 - J_3^2 + J_3.$$

Here j is $\frac{1}{2}$ and k can be either $-\frac{1}{2}$ or $\frac{1}{2}$. How big can n and n' be? Thus for what value of n do you know the quantities represented by $(J_-)^{n+1}(J_+)^{n+1}$ and $(J_+)^{n+1}(J_-)^{n+1}$ are zero regardless of what k is? Do the matrices you found agree with that?

20-6. Suppose the quantity represented by $L_1^2 + L_2^2 + L_3^2$ has the value $2\hbar^2$, that $\langle L_3 \rangle$ is 0, and that $\langle L_3^2 \rangle$ is $\hbar^2/2$. What are the possible values for the quantity represented by L_3? What are the probabilities for these values?

20-7. Use the result of Problem 17-4 to show that if $\langle L_1^2 + L_2^2 + L_3^2 \rangle$ is 0 then each of the quantities represented by L_1, L_2, L_3 has the value 0. This happens when the value $j(j+1)\hbar^2$ of the quantity represented by $L_1^2 + L_2^2 + L_3^2$ is 0.

21 ROTATIONAL ENERGY

As soon as the quantum mechanics of angular momentum were worked out by Born, Heisenberg, and Jordan, applications were made to the rotational energy of molecules [1]. They are easy to understand. The first step is to see how the rotational energy is related to the angular momentum.

For a molecule that consists of two atoms, the energy of motion of the atoms in the molecule is

$$\tfrac{1}{2}mv^2,$$

where v is the relative velocity of the atoms and

$$m = \frac{m_1 m_2}{m_1 + m_2}$$

with m_1 and m_2 being the masses of the two atoms. This does not include the energy of motion of the molecule as a whole. We may suppose the molecule is not moving. This formula is obtained the same way the formula for the energy of motion of an electron and nucleus in an atom was obtained in Chapter 19.

Suppose the only motion is rotation of the molecule around an axis through its center of mass. Then the relative position vector for the two atoms is perpendicular to the relative velocity vector, so the magnitude of the orbital angular momentum made from the relative position and momentum is

$$l = rmv,$$

where r is the distance between the atoms. Then the energy of motion of the atoms in the molecule is

$$\tfrac{1}{2}mv^2 = \frac{l^2}{2mr^2}.$$

In quantum mechanics, the energy of this rotational motion is represented by the matrix

$$\frac{1}{2\eta}\left(L_1^{\,2} + L_2^{\,2} + L_3^{\,2}\right),$$

where η is mr^2, which is a fixed number representing the "moment of inertia" of the molecule corresponding to a fixed distance r between the atoms, and L_1, L_2, L_3 are the orbital angular-momentum matrices made from the relative position and momentum matrices that we considered in the last chapter.

Here, to calculate the rotational energy, we do not use a complete model of the molecule. Our model is a molecule in which the distance between the atoms is fixed. It is a simpler physical system than a molecule in which the distance varies.

The possible values of the rotational energy are determined by those of the angular momentum. The rotational energy can have values

$$\varepsilon_j = \frac{\hbar^2}{2\eta} j(j+1) \qquad \text{for} \quad j = 0, 1, 2, 3, \ldots .$$

A molecule can emit a quantum of radiation of frequency ν when its energy changes by

$$h\nu = \varepsilon_i - \varepsilon_f,$$

where ε_i and ε_f are two of the possible values ε_j for the rotational energy. It is most likely that the energy changes to the next possible value, so the frequencies of most of the radiation should correspond to the differences

$$\begin{aligned} h\nu &= \frac{\hbar^2}{2\eta}(j+1)(j+1+1) - \frac{\hbar^2}{2\eta} j(j+1) \\ &= \frac{\hbar^2}{\eta}(j+1) \end{aligned}$$

between the energies for $j + 1$ and j, where j can have values $0, 1, 2, 3, \ldots$.

These frequencies are integral multiples of the lowest frequency corresponding to

$$h\nu = \frac{\hbar^2}{\eta} \qquad \text{for} \quad j = 0.$$

This kind of spectrum is observed. The differences between the rotational energy levels are small compared to the differences between the possible values for the energy of oscillating motion, so this model is accurate when the energy is low enough that the energy of oscillating motion can have only the lowest possible value. A more complete description is obtained by combining rotation and oscillation [2].

REFERENCES

1. J. Mehra and H. Rechenberg, *The Historical Development of Quantum Theory*, Volume 3, *The Formulation of Matrix Mechanics and Its Modifications 1925–1926*. Springer-Verlag, New York, 1982, pp. 187–193.
2. A. Böhm, *Quantum Mechanics*. Springer-Verlag, New York, 1979, Chap. III.

22 THE HYDROGEN ATOM

In quantum mechanics the energy in an atom that has one electron, for example a hydrogen atom or an ionized helium atom, is represented by a matrix

$$H = \frac{1}{2m}\left(P_1^{\,2} + P_2^{\,2} + P_3^{\,2}\right) - Ze^2R^{-1},$$

where P_1, P_2, P_3 are the matrices that represent the relative momentum of the electron and nucleus and R is a matrix that represents the distance between the electron and nucleus. The formula for H is the same as the formula for the energy used in the Bohr model, except now it is written in terms of matrices. As before, Ze is the charge of the nucleus, $-e$ is the electron charge, and

$$m = \frac{m_n m_e}{m_n + m_e}$$

where m_n is the mass of the nucleus and m_e is the electron mass.

What are the possible values of the energy represented by this matrix H? That is the problem the creators of quantum mechanics were most eager to solve. It is more difficult than the oscillator problem. Heisenberg was wise to do the oscillator first. The hydrogen problem was not solved for several months after quantum mechanics was developed, and then it was solved only by a rather indirect method.

Wolfgang Pauli was often critical. Heisenberg generally appreciated Pauli's criticism, relied on it, and sought it. It was part of the relation that began when they were students of Sommerfeld in Munich, and Pauli, who was a year and a half older, helped Heisenberg get started. They became good friends in spite of the striking differences in their personalities and life styles [1]. Heisenberg described his initial work on quantum mechanics to

Pauli in a personal visit and several letters and kept Pauli informed when quantum mechanics was developed by Heisenberg, Born, and Jordan. Pauli encouraged Heisenberg but was sharply critical of Born and the use of noncommuting matrices. Heisenberg did not appreciate that. He wrote a stinging letter to Pauli, calling his criticism a "pigsty" and a "scandal" and saying, "When you reproach us that we are such big donkeys that we have never produced anything new in physics, it may well be true. But then, you are also an equally big jackass because you have not accomplished it either." That was a challenge. In three weeks Pauli sent Heisenberg a solution of the hydrogen problem [2].

Within a few months Erwin Schrödinger described a completely different way to do quantum mechanics, using wave functions instead of matrices, and solved the hydrogen problem that way. The two ways of doing quantum mechanics are equivalent. Now everyone learns Schrödinger's method of solving the hydrogen problem. Pauli's method is easier to describe here.

Let Q_1, Q_2, Q_3 be the matrices that represent the position of the electron relative to the nucleus. Then

$$R^2 = Q_1{}^2 + Q_2{}^2 + Q_3{}^2.$$

We considered the relative position and momentum when we discussed orbital angular momentum. We showed that the matrices Q_j and P_j satisfy the same commutation relations as the position and momentum matrices for a particle.

Let L_1, L_2, L_3 be the orbital angular-momentum matrices made from Q_1, Q_2, Q_3 and P_1, P_2, P_3. They satisfy angular-momentum commutation relations.

Pauli considered the energy, angular momentum, and another vector quantity that had been used to solve problems in Newtonian mechanics where the gravity force is analogous to the Coulomb force in a hydrogen atom. It is called the Runge–Lenz vector. In quantum mechanics it is represented by the matrices

$$A_1 = \frac{1}{mZe^2}\frac{1}{2}[L_2P_3 + P_3L_2 - L_3P_2 - P_2L_3] + Q_1R^{-1},$$

$$A_2 = \frac{1}{mZe^2}\frac{1}{2}[L_3P_1 + P_1L_3 - L_1P_3 - P_3L_1] + Q_2R^{-1},$$

$$A_3 = \frac{1}{mZe^2}\frac{1}{2}[L_1P_2 + P_2L_1 - L_2P_1 - P_1L_2] + Q_3R^{-1}.$$

They have several properties that are used to solve the hydrogen problem. It can be shown, as Pauli did [3], that

$$L_1A_1 + L_2A_2 + L_3A_3 = 0,$$

that

$$A_1^2 + A_2^2 + A_3^2 = \frac{2}{mZ^2e^4}H\left(L_1^2 + L_2^2 + L_3^2 + \hbar^2\right) + 1,$$

that A_1, A_2, A_3 commute with H, that with L_1, L_2, L_3 they satisfy the commutation relations

$$L_1A_2 - A_2L_1 = i\hbar A_3, \qquad L_2A_1 - A_1L_2 = -i\hbar A_3,$$

$$L_2A_3 - A_3L_2 = i\hbar A_1, \qquad L_3A_2 - A_2L_3 = -i\hbar A_1,$$

$$L_3A_1 - A_1L_3 = i\hbar A_2, \qquad L_1A_3 - A_3L_1 = -i\hbar A_2,$$

$$L_1A_1 - A_1L_1 = 0, \qquad L_2A_2 - A_2L_2 = 0, \qquad L_3A_3 - A_3L_3 = 0,$$

and that

$$A_1A_2 - A_2A_1 = -i\frac{2\hbar}{mZ^2e^4}HL_3,$$

$$A_2A_3 - A_3A_2 = -i\frac{2\hbar}{mZ^2e^4}HL_1,$$

$$A_3A_1 - A_1A_3 = -i\frac{2\hbar}{mZ^2e^4}HL_2.$$

This requires some algebraic manipulations that involve the commutation relations for the position and momentum matrices, for the angular-momentum matrices, and for the angular-momentum matrices with the position and momentum matrices. The latter were worked out in Problem 20-2. These manipulations also involve two things we have not learned here. One is the calculation of

$$P_jR^{-1} - R^{-1}P_j = i\hbar Q_j(R^3)^{-1}$$

for $j = 1, 2, 3$. The other is that R commutes with Q_1, Q_2, Q_3 and with L_1, L_2, L_3. This implies that R^{-1} commutes with Q_1, Q_2, Q_3 and L_1, L_2, L_3.

For example, from

$$L_3 R = R L_3$$

it follows that

$$R^{-1} L_3 R R^{-1} = R^{-1} R L_3 R^{-1},$$

so

$$R^{-1} L_3 = L_3 R^{-1}.$$

We can see that R^2 commutes with Q_1, Q_2, Q_3. It was shown in Problem 20-3 that R^2 and $P_1{}^2 + P_2{}^2 + P_3{}^2$ commute with L_3. We could show the same for L_1 and L_2. It can be shown that R commutes with Q_1, Q_2, Q_3 and L_1, L_2, L_3 because R can be obtained as the square root of R^2. For 2×2 matrices, this property of the square root was shown in Problem 12-4.

The angular-momentum matrices L_1, L_2, L_3 commute with H, because they commute with $P_1{}^2 + P_2{}^2 + P_3{}^2$ and R^{-1}.

Once these properties of H, L_1, L_2, L_3 and A_1, A_2, A_3 are established, the problem can be made quite simple. The matrices L_1, L_2, L_3 and A_1, A_2, A_3 commute with H. That means they represent quantities that can be measured when the energy has a definite value. They do not all commute with each other, so all the quantities they represent generally do not have definite values together when the energy has a definite value. However, there are different possible values these quantities can have with the same value of the energy. Measuring the energy alone does not determine the state of the atom. These other quantities have to be measured too.

Suppose the energy has a certain value. We can consider the physical system consisting of an atom that has this energy. With the energy fixed, it is a simpler system than an atom where the energy can have different values. It is described by quantities that can be measured when the energy has a definite value. The quantities represented by L_1, L_2, L_3 and A_1, A_2, A_3 can be used to describe this simpler system. The relative position and momentum cannot; they generally cannot be measured when the energy has a definite value; the matrices Q_1, Q_2, Q_3 and P_1, P_2, P_3 do not commute with H.

Now we consider only the simpler system. Let ε be the energy value. This is the only value the energy can have, so the energy is represented by the matrix $\varepsilon \cdot 1$, or the number ε. The energy is a fixed number, like m or $-e$. Quantities that are not fixed numbers are represented by matrices. These include L_1, L_2, L_3 and A_1, A_2, A_3. They satisfy the same equations as they did for the more complicated system, except the matrix H in these equations is replaced by the energy value ε. This makes no difference for the algebra of L_1, L_2, L_3 and A_1, A_2, A_3 because H commuted with them for the more complicated system just as ε does for the simpler system. Thus, in the next steps, L_1, L_2, L_3 and A_1, A_2, A_3 represent quantities that describe the simpler system, and H is replaced by ε.

We made the same kind of simplification in the last chapter when we considered a molecule in which the distance between the atoms has a fixed value. It is described by quantities that can be measured when the distance has a definite value. They are represented by matrices such as L_1, L_2, L_3 that commute with the matrix R that represents the distance.

For the atom with fixed energy ε, let

$$K_1 = \sqrt{-\frac{mZ^2e^4}{2\varepsilon}}\,A_1,$$

$$K_2 = \sqrt{-\frac{mZ^2e^4}{2\varepsilon}}\,A_2,$$

$$K_3 = \sqrt{-\frac{mZ^2e^4}{2\varepsilon}}\,A_3.$$

We can see that

$$L_1K_1 + L_2K_2 + L_3K_3 = 0,$$

that

$$K_1^2 + K_2^2 + K_3^2 = -\left(L_1^2 + L_2^2 + L_3^2 + \hbar^2\right) - \frac{mZ^2e^4}{2\varepsilon},$$

that K_1, K_2, K_3 satisfy the same commutation relations with L_1, L_2, L_3 as A_1, A_2, A_3 do, and that

$$K_1K_2 - K_2K_1 = i\hbar L_3,$$

$$K_2K_3 - K_3K_2 = i\hbar L_1,$$

$$K_3K_1 - K_1K_3 = i\hbar L_2.$$

To get these from the equations involving A_1, A_2, A_3, we just multiply once or twice by $\sqrt{-mZ^2e^4/2\varepsilon}$.

Let

$$M_1 = \frac{1}{2\hbar}(L_1 + K_1), \qquad N_1 = \frac{1}{2\hbar}(L_1 - K_1),$$

$$M_2 = \frac{1}{2\hbar}(L_2 + K_2), \qquad N_2 = \frac{1}{2\hbar}(L_2 - K_2),$$

$$M_3 = \frac{1}{2\hbar}(L_3 + K_3), \qquad N_3 = \frac{1}{2\hbar}(L_3 - K_3).$$

We can see that

$$M_1^2 + M_2^2 + M_3^2 - N_1^2 - N_2^2 - N_3^2 = \frac{1}{\hbar^2}(L_1K_1 + L_2K_2 + L_3K_3) = 0,$$

that

$$2\left(M_1^2 + M_2^2 + M_3^2 + N_1^2 + N_2^2 + N_3^2\right)$$

$$= \frac{1}{\hbar^2}\left(L_1^2 + L_2^2 + L_3^2 + K_1^2 + K_2^2 + K_3^2\right) = -1 - \frac{mZ^2e^4}{2\hbar^2\varepsilon},$$

that all the matrices M_1, M_2, M_3 commute with all the matrices N_1, N_2, N_3, and that

$$M_1M_2 - M_2M_1 = iM_3, \qquad N_1N_2 - N_2N_1 = iN_3,$$

$$M_2M_3 - M_3M_2 = iM_1, \qquad N_2N_3 - N_3N_2 = iN_1,$$

$$M_3M_1 - M_1M_3 = iM_2, \qquad N_3N_1 - N_1N_3 = iN_2.$$

For example,

$$M_1N_2 - N_2M_1 = \frac{1}{4\hbar^2}[L_1L_2 - L_2L_1 - (K_1K_2 - K_2K_1)$$

$$-(L_1K_2 - K_2L_1) + K_1L_2 - L_2K_1]$$

$$= \frac{1}{4\hbar^2}[i\hbar L_3 - i\hbar L_3 - i\hbar K_3 + i\hbar K_3]$$

$$= 0,$$

$$M_1M_2 - M_2M_1 = \frac{1}{4\hbar^2}[L_1L_2 - L_2L_1 + K_1K_2 - K_2K_1$$

$$+L_1K_2 - K_2L_1 + K_1L_2 - L_2K_1]$$

$$= \frac{1}{4\hbar^2}[i\hbar L_3 + i\hbar L_3 + i\hbar K_3 + i\hbar K_3]$$

$$= i\frac{1}{2\hbar}[L_3 + K_3]$$

$$= iM_3.$$

Thus we have

$$M_1^2 + M_2^2 + M_3^2 = N_1^2 + N_2^2 + N_3^2$$

and

$$4(M_1^2 + M_2^2 + M_3^2) = -1 - \frac{mZ^2e^4}{2\hbar^2\varepsilon}.$$

We expect the energy to be negative, as it is in the Bohr model, because we are considering an electron and nucleus that are bound together in the atom and cannot move apart to an infinite distance. Hence we look for negative energy values ε. When ε is negative,

$$\sqrt{-\frac{mZ^2e^4}{2\varepsilon}}$$

is real. Then K_1, K_2, K_3 represent real quantities, so M_1, M_2, M_3 and N_1, N_2, N_3 represent real quantities. The commutation relations for M_1, M_2, M_3 are the same as for matrices J_1, J_2, J_3 that represent angular momentum divided by $\hbar$. Therefore the only values

$$M_1^2 + M_2^2 + M_3^2$$

can have are

$$j(j+1),$$

where j is one of the numbers

$$0, \tfrac{1}{2}, 1, \tfrac{3}{2}, 2, \ldots .$$

Then

$$-\frac{mZ^2e^4}{2\hbar^2\varepsilon} = 4j(j+1)+1$$

$$= (2j+1)^2,$$

so

$$\varepsilon = -\frac{m(Ze^2)^2}{2\hbar^2(2j+1)^2}.$$

Let

$$n = 2j+1.$$

Then n is one of the numbers

$$1, 2, 3, 4, \ldots,$$

and the possible energy values are

$$\varepsilon_n = -\frac{m(Ze^2)^2}{2\hbar^2 n^2}.$$

These are the same as in the Bohr model.

The energy represented by H also has a continuous range of possible values extending from zero up; every positive number is a possible value. These positive energy values are realized when the atom is ionized. Then the electron and nucleus move under the influence of the Coulomb force but can move apart to an infinite distance. Here we consider only the negative energy values for an electron and nucleus that are bound together in an atom.

Suppose the energy has the value ε_n, where n is one of the numbers $1, 2, 3, 4, \ldots$. Then the quantities represented by $M_1{}^2 + M_2{}^2 + M_3{}^2$ and $N_1{}^2 + N_2{}^2 + N_3{}^2$ have the value $j(j+1)$, where $2j+1$ is n. The possible values of the quantity represented by M_3 are the numbers

$$-j, -j+1, -j+2, \ldots, j-1, j.$$

The commutation relations for N_1, N_2, N_3 also are the same as for matrices J_1, J_2, J_3 that represent angular momentum divided by $\hbar$. Therefore the possible values of the quantity represented by N_3 are also the numbers

$$-j,\ -j+1,\ -j+2,\ldots,\ j-1,\ j.$$

There are $2j+1$ possible values for the quantity represented by M_3 and $2j+1$ possible values for the quantity represented by N_3, so the number of possible pairs of values for these two quantities is

$$(2j+1)^2 = n^2.$$

For each possible pair, there is a state of the atom where the quantities represented by M_3 and N_3 have these values, so there are n^2 different states for the given energy ε_n.

Since values of orbital angular momentum are integral multiples of $\hbar$, not half-integral, the possible values of the quantity represented by $L_1^2 + L_2^2 + L_3^2$ are $l(l+1)\hbar^2$ for non-negative integers l. We have

$$L_1^2 + L_2^2 + L_3^2 + K_1^2 + K_2^2 + K_3^2 = 4\hbar^2\left(M_1^2 + M_2^2 + M_3^2\right).$$

Suppose the energy has the value ε_n, where n is $2j+1$. Since $K_1^2 + K_2^2 + K_3^2$ represents a non-negative real quantity, it follows that

$$l(l+1)\hbar^2 \leq 4\hbar^2 j(j+1),$$

so

$$l(l+1) \leq 2j(2j+2).$$

Then

$$l \leq 2j$$

because $l = 2j+1$ would give

$$l(l+1) = (2j+1)(2j+1+1) > 2j(2j+2).$$

Since $2j$ is $n-1$, the possible values for l are

$$0, 1, 2, 3, \ldots, n-1.$$

For each l the possible values of the quantity represented by L_3 are

$$-l,\ -l+1,\ -l+2,\ldots, l-1,\ l;$$

there are $2l+1$ possible values. The number of possible pairs of values for the quantities represented by $L_1^2 + L_2^2 + L_3^2$ and L_3 is

$$[2 \cdot 0 + 1] + [2 \cdot 1 + 1] + [2 \cdot 2 + 1] + \cdots + [2(n-1) + 1]$$

$$= n\tfrac{1}{2}[1 + 2(n-1) + 1] = n^2.$$

(The sum can be calculated by multiplying the number of terms, which is n, by the average of the terms. Since the terms increase by the same amount from one term to the next, the average is half the sum of the first and last terms.) For each possible pair of values, there is a state of the atom where the quantities represented by $L_1^2 + L_2^2 + L_3^2$ and L_3 have those values, so there are n^2 different states for the given energy ε_n.

The matrices M_3 and N_3 do not commute with $L_1^2 + L_2^2 + L_3^2$, so the states where the quantities represented by M_3 and N_3 have definite values are not the same as the states where the quantities represented by $L_1^2 + L_2^2 + L_3^2$ and L_3 have definite values, but either way there are n^2 different states for the given energy ε_n. Either way, we count possible values of quantities that can be measured together.

Here we have not considered the electron spin. The spin can be measured, together with the quantities we have considered, when the energy has a definite value. The spin matrices commute with H, L_1, L_2, L_3 and A_1, A_2, A_3. There are two possible values for a projection of the spin. When they are included, the number of different states for a given energy ε_n is $2n^2$.

The numbers $2n^2$ are

$$2, 8, 18, 32, \ldots .$$

These are the numbers of electrons in closed shells, which are a basic feature of the periodic table of the elements. The application to an atom with more than one electron is made by assuming the electrons in it occupy states that are similar to those of an atom with one electron. According to the Pauli exclusion principle, there cannot be more than one electron in each state.

PROBLEMS

22-1. It can be shown, as Pauli did, that

$$\tfrac{1}{2}(A_1Q_1 + Q_1A_1 + A_2Q_2 + Q_2A_2 + A_3Q_3 + Q_3A_3)$$

$$= -\frac{1}{mZe^2}\left(L_1^2 + L_2^2 + L_3^2 + \frac{3}{2}\hbar^2\right) + R.$$

Use this and the result of Problem 17-6 to find $\langle R\rangle$ for a state where each of the quantities represented by L_1, L_2, L_3 and A_1, A_2, A_3 has the value 0. Show this occurs when the energy has the value ε_n for $n = 1$; the results of Problems 17-4 and 20-7 can be used here. Compare $\langle R\rangle$ found here with the radius of the Bohr orbit found in Problem 19-1.

22-2. Show that

$$(Q_1P_1 + P_1Q_1 + Q_2P_2 + P_2Q_2 + Q_3P_3 + P_3Q_3)(P_1^2 + P_2^2 + P_3^2)$$
$$-(P_1^2 + P_2^2 + P_3^2)(Q_1P_1 + P_1Q_1 + Q_2P_2 + P_2Q_2 + Q_3P_3 + P_3Q_3)$$
$$= i\hbar 4(P_1^2 + P_2^2 + P_3^2).$$

Use the facts that R^{-1} commutes with Q_1, Q_2, Q_3 and that

$$P_jR^{-1} - R^{-1}P_j = i\hbar Q_j(R^3)^{-1}$$

for $j = 1, 2, 3$ to show that

$$(Q_1P_1 + P_1Q_1 + Q_2P_2 + P_2Q_2 + Q_3P_3 + P_3Q_3)R^{-1}$$
$$-R^{-1}(Q_1P_1 + P_1Q_1 + Q_2P_2 + P_2Q_2 + Q_3P_3 + P_3Q_3)$$
$$= i\hbar 2R^{-1}.$$

Use these to show that

$$(Q_1P_1 + P_1Q_1 + Q_2P_2 + P_2Q_2 + Q_3P_3 + P_3Q_3)H$$
$$-H(Q_1P_1 + P_1Q_1 + Q_2P_2 + P_2Q_2 + Q_3P_3 + P_3Q_3)$$
$$= i\hbar 2\left[\frac{1}{m}(P_1^2 + P_2^2 + P_3^2) - Ze^2R^{-1}\right].$$

Use that and the result of Problem 17-5 to find

$$\frac{1}{2m}\langle P_1^2 + P_2^2 + P_3^2\rangle, \quad -Ze^2\langle R^{-1}\rangle$$

and

$$\langle P_1^2 + P_2^2 + P_3^2 \rangle, \quad \langle R^{-1} \rangle$$

for each state where the energy has a value ε_n. Compare the answers with $\langle R \rangle$ found in the last problem and with the radius and momentum of the Bohr orbit found in Problem 19-1. The difference between $1/\langle R^{-1} \rangle$ and $\langle R \rangle$ when n is 1 shows there are substantial probabilities distributed over a range of possible values of the radius that is fairly large compared to $\langle R \rangle$.

REFERENCES

1. W. Heisenberg, *Physics and Beyond.* Harper and Row, New York, 1971, pp. 24–29.
2. J. Mehra and H. Rechenberg, *The Historical Development of Quantum Theory*, Volume 3, *The Formulation of Matrix Mechanics and Its Modifications 1925–1926.* Springer-Verlag, New York, 1982, Chap. IV, Secs. 4–5, and pp. 11–12.
3. An English translation of Pauli's paper is in *Sources of Quantum Mechanics*, edited by B. L. van der Waerden. Dover, New York, 1968, p. 387.

23 SPIN ROTATIONS

We have used two very different kinds of equations. Examples of one kind are the formulas for energy and orbital angular momentum in terms of position and momentum. As relations between physical quantities, these equations would be the same without quantum mechanics. The only difference is that in quantum mechanics they are written in terms of matrices.

Examples of the other kind are the commutation relations for position and momentum and for angular momentum. The formulas for

$$QP - PQ$$

and

$$J_1J_2 - J_2J_1$$

make no sense at all without quantum mechanics because they would make no sense if they were written in terms of values rather than matrices. There were no equations like these before quantum mechanics. They are completely new. What is the meaning of these new equations?

The new equations reflect the way physical quantities change when the space and time coordinates change and the way different changes are related. The matrices are used two different ways. They represent the physical quantities that describe a particular physical system at a given time. They are also used as multipliers to change the matrices that represent physical quantities to describe the system at another time or to describe the system at a different location in space, rotated to a different orientation in space, or moving at a different velocity. This is where the new equations come in.

The commutation relations for position and momentum can be understood by considering changes in location, time, and velocity. The commutation relations for angular momentum correspond to the algebra of rotations around different axes. We consider rotations first. To begin, we consider the effect of rotations on spin.

There are two equivalent ways to look at rotations. We can imagine that the physical object or system is rotated relative to fixed reference directions, or we can describe a fixed physical system with respect to rotated reference directions. We take the latter point of view. For example, suppose my reference directions are rotated relative to yours by 180° around the 3 axis. To compare our two descriptions, we can label your directions 1, 2, 3 and mine 1′, 2′, 3′. Then 1′ and 2′ are the directions opposite 1 and 2, and 3′ is the same as 3.

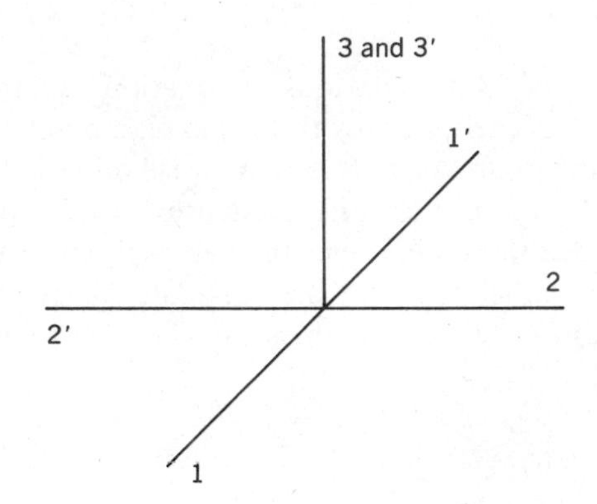

We can use Pauli matrices $\Sigma_1, \Sigma_2, \Sigma_3$ to represent the projections of a spin divided by $\hbar/2$, or a magnetic moment divided by μ, in the 1, 2, 3 directions. Then the projections in the 1′, 2′, 3′ directions are represented by the matrices

$$\Sigma_1' = -\Sigma_1,$$

$$\Sigma_2' = -\Sigma_2,$$

$$\Sigma_3' = \Sigma_3.$$

Although spin matrices are not often changed for other reasons, changes like this are fairly common. They have to be accommodated, because the choice of reference directions is often rather arbitrary; one choice may be no better than another.

This change from $\Sigma_1, \Sigma_2, \Sigma_3$ to $\Sigma_1', \Sigma_2', \Sigma_3'$ can be made by multiplying with the matrix $-i\Sigma_3$ and its inverse $i\Sigma_3$. We have

$$(-i\Sigma_3)(i\Sigma_3) = (i\Sigma_3)(-i\Sigma_3) = 1,$$

$$(i\Sigma_3)\Sigma_1(-i\Sigma_3) = -\Sigma_1,$$

$$(i\Sigma_3)\Sigma_2(-i\Sigma_3) = -\Sigma_2,$$

$$(i\Sigma_3)\Sigma_3(-i\Sigma_3) = \Sigma_3.$$

Each real quantity in the description with respect to the 1, 2, 3 directions is represented by a matrix

$$x_0 \cdot 1 + x_1\Sigma_1 + x_2\Sigma_2 + x_3\Sigma_3.$$

In the description with respect to the $1', 2', 3'$ directions there is a corresponding real quantity represented by the matrix

$$(i\Sigma_3)(x_0 \cdot 1 + x_1\Sigma_1 + x_2\Sigma_2 + x_3\Sigma_3)(-i\Sigma_3)$$

$$= x_0 \cdot 1 + x_1(-\Sigma_1) + x_2(-\Sigma_2) + x_3\Sigma_3$$

$$= x_0 \cdot 1 + x_1\Sigma_1' + x_2\Sigma_2' + x_3\Sigma_3'.$$

The matrix $-i\Sigma_3$ is the agent that changes one description to the other.

In this way, the matrix $-i\Sigma_3$ represents rotation by 180° around the 3 axis. In the same way, the matrices $-i\Sigma_1$ and $-i\Sigma_2$ represent rotations by 180° around the 1 and 2 axis.

Products of these matrices represent successive rotations. Let U and V be matrices that represent two rotations, so the matrices Σ_j for $j = 1, 2, 3$ are changed to

$$U^{-1}\Sigma_j U$$

for one rotation and to

$$V^{-1}\Sigma_j V$$

for the other. Then the two rotations in succession change the matrices Σ_j to

$$V^{-1}\left(U^{-1}\Sigma_j U\right)V = (V^{-1}U^{-1})\Sigma_j(UV)$$

$$= (UV)^{-1}\Sigma_j(UV).$$

Here we used the fact that

$$(UV)^{-1} = V^{-1}U^{-1}.$$

We see that the matrix product UV represents the rotation you get by doing one rotation and then the other. This rotation is called the product of the two rotations. Thus the product of the matrices represents the product of the rotations.

This provides an interpretation for these matrix products. For example,

$$(-i\Sigma_1)(-i\Sigma_2) = -i\Sigma_3.$$

This corresponds to the fact that the product of rotations by 180° around the 1 and 2 axes is a rotation by 180° around the 3 axis. You can verify that by rotating an object, such as a book, around three perpendicular axes.

Let U be a matrix that represents a rotation. Then U has an inverse U^{-1}. It represents the inverse rotation; from

$$(U^{-1})^{-1} = U$$

and

$$U\left(U^{-1}\Sigma_j U\right)U^{-1} = \Sigma_j$$

$$U^{-1}\left(U\Sigma_j U^{-1}\right)U = \Sigma_j$$

for $j = 1, 2, 3$ we see that the rotation represented by U^{-1} undoes the rotation represented by U, and the rotation represented by U undoes the rotation represented by U^{-1}. For example, the inverse of $-i\Sigma_3$ is $i\Sigma_3$. It represents the inverse of rotation by 180° around the 3 axis, which is another rotation by 180° around the 3 axis.

We can consider the identity rotation. It is the rotation by 0° that does not change anything. The product of a rotation and its inverse rotation is the identity rotation. The identity rotation and the three rotations by 180° around the 1, 2, 3 axes form the kind of mathematical structure that is called a group. That means the set of four rotations includes the identity rotation, includes the inverse of every rotation in the set, and includes all products of rotations in the set.

The identity rotation is represented by the matrix 1, because

$$1 \cdot \Sigma_j \cdot 1 = \Sigma_j \qquad \text{for} \quad j = 1, 2, 3.$$

From products of the matrices $-i\Sigma_1$, $-i\Sigma_2$, $-i\Sigma_3$ that represent rotations by 180° around the 1, 2, 3 axes, we get eight matrices

$$-i\Sigma_1,\ -i\Sigma_2,\ -i\Sigma_3,\ -1,$$

$$i\Sigma_1,\ i\Sigma_2,\ i\Sigma_3,\ 1.$$

These eight matrices form a group. That means they include the matrix 1, the inverse of each matrix, and all products of the matrices. The group of matrices is twice as big as the group of rotations. Each of the four rotations is represented by two matrices. For each matrix U there is a matrix $-U$ that represents the same rotation. For example, rotation by 180° around the 3 axis is represented by $i\Sigma_3$ as well as by $-i\Sigma_3$. The identity rotation is represented by both 1 and -1. Nevertheless, the products of the matrices represent the products of the rotations. Multiplication of these matrices corresponds to multiplication of rotations.

This raises some questions. There are at least two matrices for each rotation. Are there more? If so, why do we choose these particular matrices? Are there others that work equally well?

These four rotations commute with each other. The eight matrices do not commute. Why do we use products of matrices that do not commute to represent products of rotations that do?

In general, rotations do not commute. They are like matrices in that respect. We can see this by including 90° rotations. That will also provide a way to explain the matrices chosen to represent 180° rotations. There are other matrices that work equally well for 180° rotations, but they do not work equally well when 90° rotations are included. We investigate them for 180° rotations before we look at 90° rotations.

The first step is to see which matrices represent the identity rotation. Suppose G is a 2×2 matrix that represents the identity rotation. Then

$$G^{-1}\Sigma_j G = \Sigma_j,$$

so

$$\Sigma_j G = G\Sigma_j \qquad \text{for} \quad j = 1, 2, 3.$$

This implies there is a complex number z such that

$$G = z \cdot 1.$$

The matrices that represent the identity rotation are complex multiples of 1.

Now we can see how many matrices represent one rotation. Suppose U and V are matrices that represent the same rotation. Then VU^{-1} represents the identity rotation, because it is the product of matrices that represent a rotation and its inverse, and a product of a rotation and its inverse is the identity rotation. Therefore

$$VU^{-1} = z \cdot 1$$

for some complex number z. Then

$$\begin{aligned} V &= VU^{-1}U \\ &= zU. \end{aligned}$$

Conversely, for any complex number z that is not 0, the matrix zU represents the rotation as well as U, because

$$\begin{aligned} (zU)^{-1}\Sigma_j(zU) &= z^{-1}U^{-1}\Sigma_j zU \\ &= U^{-1}\Sigma_j U \qquad \text{for} \quad j = 1,2,3. \end{aligned}$$

Thus the matrices that represent the same rotation are the matrices you get by multiplying one of them with each of the complex numbers except 0.

For example, for each complex number z that is not 0, the matrix $z(-i\Sigma_3)$ represents rotation by 180° around the 3 axis.

It is nice to have a group of matrices to represent a group of rotations. For that, different matrices that represent the same rotation may not work equally well. They may make the group of matrices unnecessarily complicated. For example, suppose rotation by 180° around the 3 axis is represented by the matrix $z(-i\Sigma_3)$, where z is a complex number. The square of this matrix is $-z^2$. The square of that is $(-z^2)^2$. If $(-z^2)^2$ is not 1, then $-z^2$ is not 1 and $(-z^2)^2$ is not the same as $-z^2$. Then the group of matrices that contains $z(-i\Sigma_3)$ must contain at least three different matrices

$$z(-i\Sigma_3), \qquad -z^2z(-i\Sigma_3), \qquad (-z^2)^2z(-i\Sigma_3)$$

that represent the same rotation, because the group must contain all products of matrices in it. It is not necessary to have more than two matrices for each rotation. Let us agree not to have more. Then $(-z^2)^2$ must be 1. That implies z is one of the numbers

$$1, -1, i, -i.$$

Thus our choice of matrices that represent rotation by 180° around the 3 axis is limited to

$$-i\Sigma_3,\ i\Sigma_3,\ \Sigma_3,\ -\Sigma_3.$$

Similarly, the choice of matrices is limited to

$$-i\Sigma_1,\ i\Sigma_1,\ \Sigma_1,\ -\Sigma_1$$

for rotations by 180° around the 1 axis and to

$$-i\Sigma_2,\ i\Sigma_2,\ \Sigma_2,\ -\Sigma_2$$

for rotations by 180° around the 2 axis.

Now consider the group of these 180° rotations around the 1, 2, 3 axes and the identity rotation. What choices do we have for a group of matrices to represent this group of four rotations? It must contain a matrix $-i\Sigma$ or $i\Sigma$ for at least one of the 180° rotations; it is not possible to use only the matrices Σ and $-\Sigma$ for all three 180° rotations because

$$\Sigma_1\Sigma_2 = i\Sigma_3.$$

Then the group of matrices contains -1, because -1 is the square of both $-i\Sigma$ and $i\Sigma$. This implies that for each matrix U in the group, the matrix $-U$ also is in the group, so the group contains at least two matrices for each rotation. There are various ways to combine two matrices for each rotation to form a group. We have considered one. An alternative is considered in Problem 23-4. The combination we considered is the standard one because it is the only one that works when 90° rotations are included.

Suppose the $1', 2', 3'$ directions are rotated 90° around the 3 axis relative to the 1, 2, 3 directions, so the $1'$ direction is opposite the 2 direction, $2'$ is the same as 1, and $3'$ is the same as 3. Then

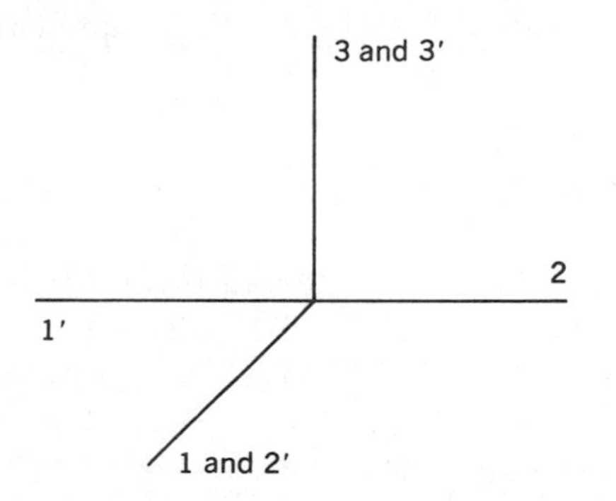

$$\Sigma_1' = -\Sigma_2,$$

$$\Sigma_2' = \Sigma_1,$$

$$\Sigma_3' = \Sigma_3.$$

This change can be made by multiplying with the matrix $(1/\sqrt{2})(1 - i\Sigma_3)$ and its inverse $(1/\sqrt{2})(1 + i\Sigma_3)$. We have

$$\frac{1}{\sqrt{2}}(1 - i\Sigma_3)\frac{1}{\sqrt{2}}(1 + i\Sigma_3) = \frac{1}{\sqrt{2}}(1 + i\Sigma_3)\frac{1}{\sqrt{2}}(1 - i\Sigma_3) = 1,$$

$$\frac{1}{\sqrt{2}}(1 + i\Sigma_3)\Sigma_1\frac{1}{\sqrt{2}}(1 - i\Sigma_3) = -\Sigma_2,$$

$$\frac{1}{\sqrt{2}}(1 + i\Sigma_3)\Sigma_2\frac{1}{\sqrt{2}}(1 - i\Sigma_3) = \Sigma_1,$$

$$\frac{1}{\sqrt{2}}(1 + i\Sigma_3)\Sigma_3\frac{1}{\sqrt{2}}(1 - i\Sigma_3) = \Sigma_3.$$

Each matrix that represents a real quantity in the description with respect to the 1, 2, 3 directions is changed to a matrix

$$\frac{1}{\sqrt{2}}(1 + i\Sigma_3)(x_0 \cdot 1 + x_1\Sigma_1 + x_2\Sigma_2 + x_3\Sigma_3)\frac{1}{\sqrt{2}}(1 - i\Sigma_3)$$

$$= x_0 \cdot 1 + x_1(-\Sigma_2) + x_2\Sigma_1 + x_3\Sigma_3$$

$$= x_0 \cdot 1 + x_1\Sigma_1' + x_2\Sigma_2' + x_3\Sigma_3',$$

which represents the corresponding quantity in the description with respect to the $1', 2', 3'$ directions. The matrix $(1/\sqrt{2})(1 - i\Sigma_3)$ is the agent that changes one description to the other.

In this way the matrix $(1/\sqrt{2})(1 - i\Sigma_3)$ represents rotation by 90° around the 3 axis. Its square is

$$\frac{1}{\sqrt{2}}(1 - i\Sigma_3)\frac{1}{\sqrt{2}}(1 - i\Sigma_3) = -i\Sigma_3,$$

which represents rotation by 180° around the 3 axis, so the square of the matrix represents the square of the 90° rotation. In the same way, the matrices $(1/\sqrt{2})(1 - i\Sigma_1)$ and $(1/\sqrt{2})(1 - i\Sigma_2)$ represent 90° rotations around the 1 and 2 axes.

Products of these matrices make a group of 48 matrices. Products of the 90° rotations around the 1, 2, 3 axes form a group of 24 rotations; there are four ways a cube can be rotated by multiples of 90° with the same face up, and there are six faces that can be up. There are two matrices for each rotation. For each matrix U in the group, there is also the matrix $-U$ that represents the same rotation.

These rotations do not commute. You can verify that by rotating an object 90° around two perpendicular axes. Some examples are considered in Problems 23-6 and 23-7. When 90° rotations are included, there is a more complete correspondence between multiplication of the matrices and multiplication of rotations.

Each 180° rotation is represented by matrices $-i\Sigma$ and $i\Sigma$. This is required for a group of matrices with just two matrices for each rotation when 90° rotations are included. We get more than two matrices for a rotation if we use the matrices Σ and $-\Sigma$ for any 180° rotation. This is demonstrated in Problem 23-8.

This representation of rotations by matrices can be extended to all rotations. The rotations form a group. A group of 2×2 matrices can be made to represent the rotations. The group of matrices is twice as big as the group of rotations. It contains matrices for every rotation, for every angle around every axis. For each rotation there are two matrices; for each matrix U in the group, there is also the matrix $-U$ that represents the same rotation. Each matrix U representing a rotation changes the matrices Σ_j to matrices $U^{-1}\Sigma_j U$ the same as the rotation does. Products of the matrices in the group represent products of the rotations. The matrices for the identity rotation and rotations by 180° and 90° are the ones we have considered.

Thus the Pauli matrices and the matrices made from them play two different roles. They represent the physical quantities that describe a spin and magnetic moment. They also represent rotations; they are used to change the matrices that represent physical quantities to describe the effects of rotations. A product of two matrices that represent rotations represents the product of the rotations, so multiplication of these matrices corresponds to multiplication of rotations. The full meaning of the algebraic relations between matrices involves both of these roles.

PROBLEMS

23-1. Let U be a matrix that has an inverse and let A, B, C and A', B', C' be matrices related by

$$A' = U^{-1}AU,$$

$$B' = U^{-1}BU,$$

$$C' = U^{-1}CU.$$

Show that if

$$A + B = C,$$

then

$$A' + B' = C'.$$

Show that if

$$AB = C,$$

then

$$A'B' = C';$$

in particular, if

$$A^2 = C,$$

then

$$(A')^2 = C'.$$

Show that if there is a complex number z such that

$$zB = C,$$

then

$$zB' = C'.$$

Show that if B has an inverse and

$$B^{-1} = C,$$

then B' has an inverse and

$$(B')^{-1} = C'.$$

These show that algebraic relations between matrices are not changed when all the matrices are changed by multiplication with U and U^{-1}.

23-2. Write out the multiplication rules for the three matrices $U^{-1}\Sigma_j U$ for $j = 1, 2, 3$, and thus show they are the same as for the three Pauli matrices Σ_j. Everything that follows from the multiplication rules is therefore the same for the matrices $U^{-1}\Sigma_j U$ as for the matrices Σ_j. For example, the possible values of the quantities the matrices represent are the same.

23-3. Consider all the products in the group of eight matrices representing the group of 180° rotations around the 1, 2, 3 axes and the identity rotation. By rotating an object, find all the products of the four rotations and verify that each matrix product represents the product of the corresponding rotations.

23-4. Show that the eight matrices

$$\Sigma_1,\ -i\Sigma_2,\ \Sigma_3,\ -1$$

$$-\Sigma_1,\ i\Sigma_2,\ -\Sigma_3,\ 1$$

form a group. This is an alternative that works as well as the standard combination to represent the group of 180° rotations around the 1, 2, 3 axes and the identity rotation. All these matrices are real.

23-5. Work out $[(1/\sqrt{2})(1 + i\Sigma_3)]^{-1}\Sigma_j(1/\sqrt{2})(1 + i\Sigma_3)$ for $j = 1, 2, 3$ and thus show that $(1/\sqrt{2})(1 + i\Sigma_3)$ represents rotation by 90° around the 3 axis in the direction opposite that of the rotation represented by $(1/\sqrt{2})(1 - i\Sigma_3)$.

23-6. Find the products of the matrices $(1/\sqrt{2})(1 - i\Sigma_1)$ and $(1/\sqrt{2})(1 - i\Sigma_2)$ that represent 90° rotations around the 1 and 2 axes. Show that the matrices do not commute and that the products of the two different orders of multiplication are matrices that represent different rotations. By rotating an object, verify that 90° rotations around two perpendicular axes do not commute.

23-7. Do the same as in the last problem for the matrices $(1/\sqrt{2})(1 - i\Sigma_1)$ and $-i\Sigma_2$ for rotations by 90° around the 1 axis and 180° around the 2 axis.

23-8. Show that the group of eight matrices considered in Problem 4 cannot be extended to include 90° rotations with just two matrices for each rotation. To do this, show that 90° rotations would be represented by matrices

$$z_1\sqrt{i}\,\frac{1}{\sqrt{2}}(1 - i\Sigma_1),\quad z_2\frac{1}{\sqrt{2}}(1 - i\Sigma_2),\quad z_3\sqrt{i}\,\frac{1}{\sqrt{2}}(1 - i\Sigma_3)$$

with the choice of each complex number z_1, z_2, z_3 limited to

1, -1, i, $-i$. Work out the products

$$z_1\sqrt{i}\,\frac{1}{\sqrt{2}}(1 - i\Sigma_1)z_2\frac{1}{\sqrt{2}}(1 - i\Sigma_2)$$

and

$$z_3\sqrt{i}\,\frac{1}{\sqrt{2}}(1 - i\Sigma_3)z_1\sqrt{i}\,\frac{1}{\sqrt{2}}(1 - i\Sigma_1).$$

Using the fact that the group of matrices contains the matrix -1, show that you get four matrices for the same rotation. It is easy to see similarly that the group of matrices will include more than two matrices for the same rotation when 90° rotations are included if the matrices Σ and $-\Sigma$ are used for any 180° rotation.

24 SMALL ROTATIONS

Originally, the Pauli matrices were constructed by assuming that the matrices representing spin angular momentum satisfy the same commutation relations as the matrices representing orbital angular momentum. These commutation relations were obtained from the formulas for orbital angular momentum in terms of position and momentum and the commutation relations for the matrices representing position and momentum. The spin matrices were described by Pauli in a paper published in 1927. In the same year John von Neumann and Eugene Wigner showed that the commutation relations for angular momentum correspond to characteristic properties of matrices that represent the group of rotations [1]. This gives them a fundamental meaning that is independent of position and momentum.

Wigner recalls that this seemed "very natural to observe" [2]. He looked at things in terms of their symmetries. He was trained as a chemical engineer and had studied symmetries of crystal structure. When he read the quantum mechanics developed by Heisenberg, Born, and Jordan, he was "overwhelmed" with "how wonderful it is" and began to seek its deeper meaning [2]. Wigner and von Neumann had been friends since high school in Hungary, and Wigner turned to his old friend when he wanted help with mathematical problems. In this case, von Neumann provided knowledge of work that had already been done on representations of groups by matrices.

To see how angular-momentum commutation relations correspond to multiplication of rotations, we consider small rotations. Every rotation can be made by multiplying a number of small rotations together. This is how the group of all rotations is usually studied. It leads rather directly to the commutation relations for angular momentum. We shall also see how rotations of physical quantities are characterized by commutation relations between the matrices that represent those quantities and the matrices for angular momentum.

In this approach we restrict our attention to small rotations but extend it to include various physical systems. We consider the general case of a system that can be rotated without its description being changed any more than by rotation of arbitrary reference directions. We have studied 180° and 90° rotations for a spin and magnetic moment as an example that shows how the description of the rotated system is obtained by multiplying the matrices that represent physical quantities with matrices that represent rotations. The general case can be handled the same way.

The first step is to describe small rotations. Consider a rotation through a small angle around the 3 axis. Suppose a vector of length x_1 in the 1 direction is rotated to a vector that has a projection εx_1 in the 2 direction, where ε is a small positive number. This is shown in the first two drawings. The rotation does not change the length of the vector. Thus the projection of the rotated vector in the 1 direction is approximately

$$\left(1 - \tfrac{1}{2}\varepsilon^2\right)x_1$$

because this gives approximately

$$\left[\left(1 - \tfrac{1}{2}\varepsilon^2\right)x_1\right]^2 + \left[\varepsilon x_1\right]^2 = x_1^{\,2}.$$

To make this approximation, we ignore $\varepsilon^4 x_1^{\,2}/4$. In working with these small rotations, we always keep terms with ε and ε^2 and ignore terms with $\varepsilon^3, \varepsilon^4, \ldots$. If ε is small, then ε^2 is ε times smaller, and ε^3 is ε times smaller than ε^2. For example, if ε is .01, then ε^2 is .0001 and ε^3 is .000001.

The same rotation takes a vector of length x_2 in the 2 direction to a vector with projections $-\varepsilon x_2$ in the 1 direction and $(1 - \varepsilon^2/2)x_2$ in the 2 direction. That is shown in the second two drawings. By adding, we see that a vector with projections x_1 and x_2 in the 1 and 2 directions is rotated to a vector that has projections

$$\left(1 - \tfrac{1}{2}\varepsilon^2\right)x_1 - \varepsilon x_2$$

in the 1 direction and

$$\varepsilon x_1 + \left(1 - \tfrac{1}{2}\varepsilon^2\right)x_2$$

in the 2 direction. That is shown in the last two drawings. If the vector also has a projection x_3 in the 3 direction, that projection is not changed by this rotation around the 3 axis.

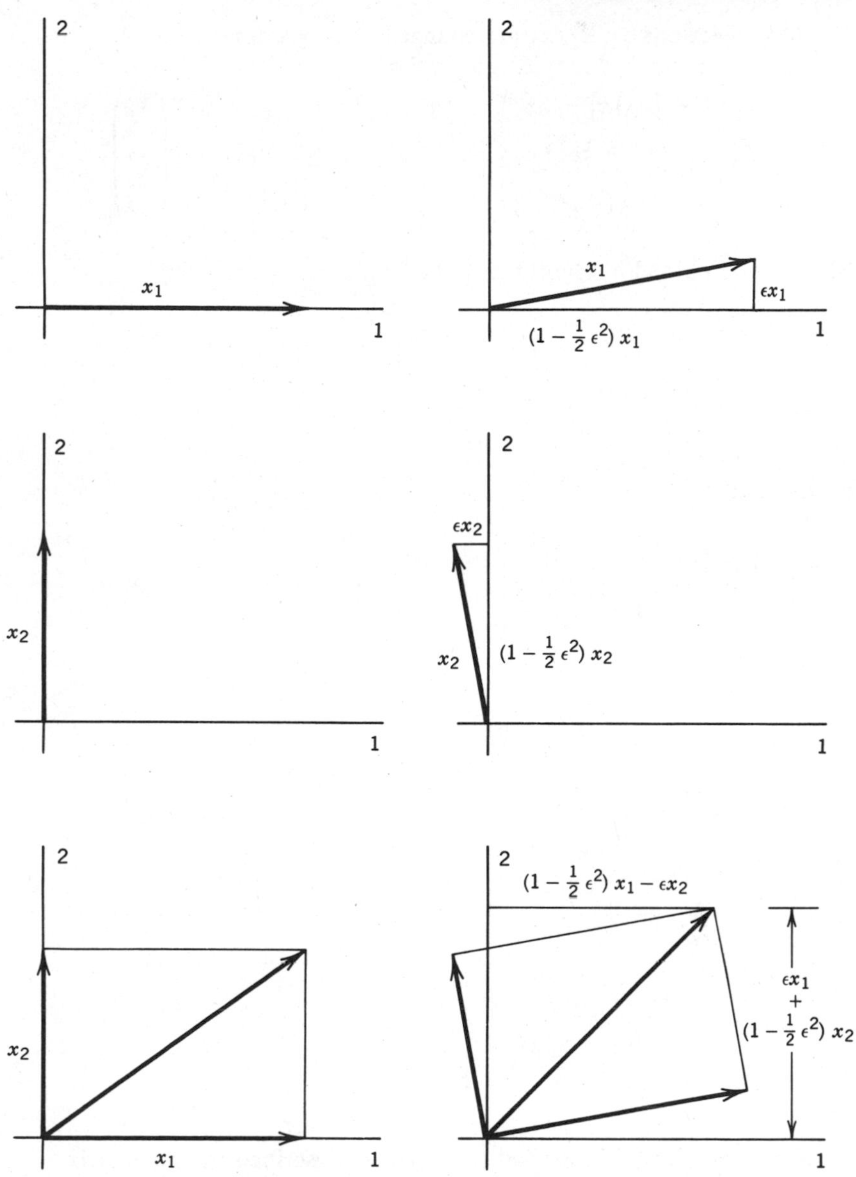

We get the same changes for the projections if we keep the vector fixed and rotate the reference directions by the same angle in the opposite direction around the 3 axis. That is the point of view we used to describe 180° and 90° rotations for a spin and magnetic moment. You can look at it either way.

The projections x_1, x_2, x_3 are changed to projections

$$\begin{pmatrix} (1 - \frac{1}{2}\varepsilon^2)x_1 - \varepsilon x_2 \\ \varepsilon x_1 + (1 - \frac{1}{2}\varepsilon^2)x_2 \\ x_3 \end{pmatrix} = \begin{pmatrix} 1 - \frac{1}{2}\varepsilon^2 & -\varepsilon & 0 \\ \varepsilon & 1 - \frac{1}{2}\varepsilon^2 & 0 \\ 0 & 0 & 1 \end{pmatrix} \begin{pmatrix} x_1 \\ x_2 \\ x_3 \end{pmatrix},$$

which are obtained by multiplying the single-column matrix

$$\begin{pmatrix} x_1 \\ x_2 \\ x_3 \end{pmatrix}$$

by the matrix

$$\begin{pmatrix} 1 - \frac{1}{2}\varepsilon^2 & -\varepsilon & 0 \\ \varepsilon & 1 - \frac{1}{2}\varepsilon^2 & 0 \\ 0 & 0 & 1 \end{pmatrix}$$

$$= \begin{pmatrix} 1 & 0 & 0 \\ 0 & 1 & 0 \\ 0 & 0 & 1 \end{pmatrix} - i\varepsilon \begin{pmatrix} 0 & -i & 0 \\ i & 0 & 0 \\ 0 & 0 & 0 \end{pmatrix} - \tfrac{1}{2}\varepsilon^2 \begin{pmatrix} 1 & 0 & 0 \\ 0 & 1 & 0 \\ 0 & 0 & 0 \end{pmatrix}$$

$$= 1 - i\varepsilon M_3 - \tfrac{1}{2}\varepsilon^2 M_3^{\,2},$$

where

$$M_3 = \begin{pmatrix} 0 & -i & 0 \\ i & 0 & 0 \\ 0 & 0 & 0 \end{pmatrix}.$$

Thus the small rotation around the 3 axis is described by this matrix.

This matrix should not be confused with either the matrices that represent physical quantities or the matrices that multiply them to represent rotations. Those matrices are usually not 3×3 like this one. They are characteristic of the physical system. This matrix does not refer to a specific system. It describes how the projections of any vector are changed by this rotation.

Small rotations around the 1 and 2 axes can be described similarly with the matrices

$$\begin{pmatrix} 1 & 0 & 0 \\ 0 & 1-\frac{1}{2}\varepsilon^2 & -\varepsilon \\ 0 & \varepsilon & 1-\frac{1}{2}\varepsilon^2 \end{pmatrix}$$

$$= \begin{pmatrix} 1 & 0 & 0 \\ 0 & 1 & 0 \\ 0 & 0 & 1 \end{pmatrix} - i\varepsilon \begin{pmatrix} 0 & 0 & 0 \\ 0 & 0 & -i \\ 0 & i & 0 \end{pmatrix} - \tfrac{1}{2}\varepsilon^2 \begin{pmatrix} 0 & 0 & 0 \\ 0 & 1 & 0 \\ 0 & 0 & 1 \end{pmatrix}$$

$$= 1 - i\varepsilon M_1 - \tfrac{1}{2}\varepsilon^2 M_1{}^2$$

and

$$\begin{pmatrix} 1-\frac{1}{2}\varepsilon^2 & 0 & \varepsilon \\ 0 & 1 & 0 \\ -\varepsilon & 0 & 1-\frac{1}{2}\varepsilon^2 \end{pmatrix}$$

$$= \begin{pmatrix} 1 & 0 & 0 \\ 0 & 1 & 0 \\ 0 & 0 & 1 \end{pmatrix} - i\varepsilon \begin{pmatrix} 0 & 0 & i \\ 0 & 0 & 0 \\ -i & 0 & 0 \end{pmatrix} - \tfrac{1}{2}\varepsilon^2 \begin{pmatrix} 1 & 0 & 0 \\ 0 & 0 & 0 \\ 0 & 0 & 1 \end{pmatrix}$$

$$= 1 - i\varepsilon M_2 - \tfrac{1}{2}\varepsilon^2 M_2{}^2,$$

where

$$M_1 = \begin{pmatrix} 0 & 0 & 0 \\ 0 & 0 & -i \\ 0 & i & 0 \end{pmatrix} \quad \text{and} \quad M_2 = \begin{pmatrix} 0 & 0 & i \\ 0 & 0 & 0 \\ -i & 0 & 0 \end{pmatrix}.$$

In each case, changing ε to $-\varepsilon$ gives the inverse rotation. It is described by the matrix

$$1 + i\varepsilon M - \tfrac{1}{2}\varepsilon^2 M^2$$

because in this approximation

$$\begin{aligned}(1 + i\varepsilon M - \tfrac{1}{2}\varepsilon^2 M^2)(1 - i\varepsilon M - \tfrac{1}{2}\varepsilon^2 M^2) &= 1 + \varepsilon^2 M^2 - \tfrac{1}{2}\varepsilon^2 M^2 - \tfrac{1}{2}\varepsilon^2 M^2 \\ &= 1,\end{aligned}$$

and similarly

$$(1 - i\varepsilon M - \tfrac{1}{2}\varepsilon^2 M^2)(1 + i\varepsilon M - \tfrac{1}{2}\varepsilon^2 M^2) = 1.$$

These matrices just describe the rotations. Now we consider the effect of the rotations on the description of a system in quantum mechanics.

Corresponding to the small rotation around the 3 axis, there are matrices U and U^{-1} that multiply the matrices representing physical quantities so the effect of the rotation is described by changing each matrix B that represents a physical quantity to $U^{-1}BU$. Moreover, there is a matrix J_3 such that for any small real number ε

$$U = 1 - i\varepsilon J_3$$

in the first approximation where terms with ε are included but terms with $\varepsilon^2, \varepsilon^3, \ldots$ are ignored. In the same approximation

$$U^{-1} = 1 + i\varepsilon J_3$$

because

$$(1 + i\varepsilon J_3)(1 - i\varepsilon J_3) = 1$$

and

$$(1 - i\varepsilon J_3)(1 + i\varepsilon J_3) = 1.$$

In this approximation

$$\begin{aligned} U^{-1}BU &= (1 + i\varepsilon J_3)B(1 - i\varepsilon J_3) \\ &= B - i\varepsilon(BJ_3 - J_3B). \end{aligned}$$

For the identity rotation, ε is 0, so U is 1 and the matrices B that represent physical quantities are not changed. For small rotations, when ε is small but not zero, the change in a matrix B that represents a physical quantity is proportional to ε; the amount of change is proportional to the amount of rotation. The i is included so J_3 can represent a real quantity; if J_3 and B represent real quantities, the general rules imply that

$$-i(BJ_3 - J_3B)$$

represents a real quantity, so the matrix B is changed to a matrix $U^{-1}BU$ that represents a real quantity. All this explains the form of U in terms of J_3 and ε in this approximation.

To include terms with ε^2, we write

$$U = 1 - i\varepsilon J_3 - \tfrac{1}{2}\varepsilon^2 K,$$

where K is whatever matrix we need in the next approximation where we keep terms with ε and ε^2 but ignore those with $\varepsilon^3, \varepsilon^4, \ldots$. Changing ε to $-\varepsilon$ gives the inverse rotation, so we should have

$$U^{-1} = 1 + i\varepsilon J_3 - \tfrac{1}{2}\varepsilon^2 K.$$

That means

$$\left(1 + i\varepsilon J_3 - \tfrac{1}{2}\varepsilon^2 K\right)\left(1 - i\varepsilon J_3 - \tfrac{1}{2}\varepsilon^2 K\right) = 1$$

or

$$1 + \varepsilon^2 J_3^2 - \tfrac{1}{2}\varepsilon^2 K - \tfrac{1}{2}\varepsilon^2 K = 1,$$

which implies

$$K = J_3^2.$$

Thus for any small real number ε we have

$$U = 1 - i\varepsilon J_3 - \tfrac{1}{2}\varepsilon^2 J_3^2$$

and

$$U^{-1} = 1 + i\varepsilon J_3 - \tfrac{1}{2}\varepsilon^2 J_3^2.$$

Similarly, there are matrices J_1 and J_2 such that for any small real number ε

$$1 - i\varepsilon J_1 - \tfrac{1}{2}\varepsilon^2 J_1^2$$

and

$$1 - i\varepsilon J_2 - \tfrac{1}{2}\varepsilon^2 J_2^2$$

correspond to small rotations around the 1 and 2 axes. Products of these matrices for rotations around the 1, 2, 3 axes correspond to products of the rotations. For any of these rotations, a matrix B that represents a physical quantity is changed to $U^{-1}BU$, where U is the matrix corresponding to the rotation.

Rotations are generally represented this way in quantum mechanics. The whole scheme follows from quite simple assumptions [3–9]. It is assumed that the system can be rotated without its description being changed any more than by rotation of arbitrary reference directions, so the descriptions before and after the rotation are equivalent. It is assumed that the way the description changes does not depend on the initial orientation of the system but depends only on the change in orientation produced by the rotation, so after one rotation, a second rotation is represented the same way; the representation of the second rotation does not depend on the orientation of the system produced by the first. This assumption that the orientation makes no difference is justified for an isolated system, because an isolated system cannot be oriented relative to anything that influences it.

From these and minor technical assumptions, it can be shown that for each axis there is a matrix J such that the matrices

$$U = 1 - i\varepsilon J - \tfrac{1}{2}\varepsilon^2 J^2$$

represent the small rotations around that axis as agents for the change between the matrices B and $U^{-1}BU$ that represent physical quantities. It can be shown that products of these matrices U represent products of the rotations around different axes. When only small rotations are considered, there is just one matrix U for each rotation. Since we are not making a complete group of matrices, we have no need for two matrices U and $-U$ for the same rotation. In the group of matrices that represent rotations for a spin and magnetic moment, the matrix -1 is included to represent 360° rotations obtained as squares of 180° rotations. Here we do not consider such big rotations. The identity rotation is represented by the matrix 1, not -1, and every small rotation is represented by a matrix that differs from 1 by small terms involving $\varepsilon, \varepsilon^2, \ldots$.

In the approximation where we include only terms with ε and ε^2, we find that

$$\begin{aligned}
&\left(1 + i\varepsilon J_2 - \tfrac{1}{2}\varepsilon^2 J_2{}^2\right)\left(1 + i\varepsilon J_1 - \tfrac{1}{2}\varepsilon^2 J_1{}^2\right)\left(1 - i\varepsilon J_2 - \tfrac{1}{2}\varepsilon^2 J_2{}^2\right) \\
&\quad\times\left(1 - i\varepsilon J_1 - \tfrac{1}{2}\varepsilon^2 J_1{}^2\right) = 1 + i\varepsilon J_2 + i\varepsilon J_1 - i\varepsilon J_2 - i\varepsilon J_1 \\
&\qquad + i\varepsilon J_2(-i\varepsilon J_2) - \tfrac{1}{2}\varepsilon^2 J_2{}^2 - \tfrac{1}{2}\varepsilon^2 J_2{}^2 \\
&\qquad + i\varepsilon J_1(-i\varepsilon J_1) - \tfrac{1}{2}\varepsilon^2 J_1{}^2 - \tfrac{1}{2}\varepsilon^2 J_1{}^2 + i\varepsilon J_2 i\varepsilon J_1 \\
&\qquad + i\varepsilon J_2(-i\varepsilon J_1) + i\varepsilon J_1(-i\varepsilon J_2) - i\varepsilon J_2(-i\varepsilon J_1) \\
&\qquad = 1 + \varepsilon^2(J_1 J_2 - J_2 J_1).
\end{aligned}$$

For any small real number ε, this product of four matrices represents the product of four rotations described by

$$(1 + i\varepsilon M_2 - \tfrac{1}{2}\varepsilon^2 M_2{}^2)(1 + i\varepsilon M_1 - \tfrac{1}{2}\varepsilon^2 M_1{}^2)(1 - i\varepsilon M_2 - \tfrac{1}{2}\varepsilon^2 M_2{}^2)$$

$$\times(1 - i\varepsilon M_1 - \tfrac{1}{2}\varepsilon^2 M_1{}^2) = 1 + \varepsilon^2(M_1 M_2 - M_2 M_1).$$

The reason the orders are the same rather than opposite is shown in Problem 24-7. By multiplying the matrices M_1 and M_2 and subtracting the products, we find that

$$M_1 M_2 - M_2 M_1 = iM_3.$$

Therefore the product of the four rotations is described by

$$1 + i\varepsilon^2 M_3.$$

It is a rotation around the 3 axis. It takes a vector of length x_1 in the 1 direction to a vector that has a projection $-\varepsilon^2 x_1$ in the 2 direction. Then

$$1 + \varepsilon^2(J_1 J_2 - J_2 J_1) = 1 + i\varepsilon^2 J_3$$

because the product of the four matrices is the matrix that represents the product of the four rotations; this is part of the general scheme that follows from assumptions valid for any isolated system. The last equation is not an approximation. It has to hold when higher powers of ε are included because the full equation holds for any choice of the small real number ε. Therefore

$$J_1 J_2 - J_2 J_1 = iJ_3.$$

It follows similarly that

$$J_2 J_3 - J_3 J_2 = iJ_1,$$

$$J_3 J_1 - J_1 J_3 = iJ_2.$$

These are the characteristic commutation relations for the matrices J_1, J_2, J_3 corresponding to rotations. They are also characteristic of angular momentum. Thus the commutation relations for angular momentum correspond to multiplication of rotations.

The matrices J_1, J_2, J_3 generally play two roles. The matrices U made from J_1, J_2, J_3 are used to describe rotations by changing the matrices B

that represent physical quantities to $U^{-1}BU$. In addition, $\hbar J_1, \hbar J_2, \hbar J_3$ are the matrices that represent the angular momentum. This will be demonstrated more specifically by examples.

Now we consider particular ways matrices are changed by rotations. They are characteristic of the physical quantities that describe the system. For example, consider a vector quantity with projections represented by matrices A_1, A_2, A_3. We know how vector projections x_1, x_2, x_3 are changed by rotations. The matrices A_1, A_2, A_3 must be changed the same way. For a rotation around the 3 axis for any small real number ε, we must have

$$\left(1 + i\varepsilon J_3 - \tfrac{1}{2}\varepsilon^2 J_3{}^2\right) A_1 \left(1 - i\varepsilon J_3 - \tfrac{1}{2}\varepsilon^2 J_3{}^2\right) = \left(1 - \tfrac{1}{2}\varepsilon^2\right) A_1 - \varepsilon A_2,$$

$$\left(1 + i\varepsilon J_3 - \tfrac{1}{2}\varepsilon^2 J_3{}^2\right) A_2 \left(1 - i\varepsilon J_3 - \tfrac{1}{2}\varepsilon^2 J_3{}^2\right) = \varepsilon A_1 + \left(1 - \tfrac{1}{2}\varepsilon^2\right) A_2,$$

$$\left(1 + i\varepsilon J_3 - \tfrac{1}{2}\varepsilon^2 J_3{}^2\right) A_3 \left(1 - i\varepsilon J_3 - \tfrac{1}{2}\varepsilon^2 J_3{}^2\right) = A_3.$$

When we work out the products on the left, keep the terms with ε, and ignore those with $\varepsilon^2, \varepsilon^3, \ldots,$ we get

$$A_1 - i\varepsilon\left(A_1 J_3 - J_3 A_1\right) = A_1 - \varepsilon A_2,$$

$$A_2 - i\varepsilon\left(A_2 J_3 - J_3 A_2\right) = \varepsilon A_1 + A_2,$$

$$A_3 - i\varepsilon\left(A_3 J_3 - J_3 A_3\right) = A_3.$$

These equations are not approximations. They have to hold when higher powers of ε are included for the full equations to hold for any choice of the small number ε. Therefore

$$A_1 J_3 - J_3 A_1 = -iA_2,$$

$$A_2 J_3 - J_3 A_2 = iA_1,$$

$$A_3 J_3 - J_3 A_3 = 0.$$

For rotations around the 1 and 2 axes, we get similar equations with J_1 and J_2 instead of J_3. These commutation relations characterize rotations of a vector quantity. We have seen how they are obtained from the formulas for the projections of the rotated vector. Conversely, those formulas are implied by these simpler commutation relations; using the latter, we get

$$
\begin{aligned}
&(1 + i\varepsilon J_3 - \tfrac{1}{2}\varepsilon^2 J_3{}^2)A_1(1 - i\varepsilon J_3 - \tfrac{1}{2}\varepsilon^2 J_3{}^2)\\
&\quad = A_1 - i\varepsilon(A_1J_3 - J_3A_1) - \tfrac{1}{2}\varepsilon^2[(A_1J_3 - J_3A_1)J_3 - J_3(A_1J_3 - J_3A_1)]\\
&\quad = A_1 - i\varepsilon(-iA_2) - \tfrac{1}{2}\varepsilon^2(-i)(A_2J_3 - J_3A_2)\\
&\quad = A_1 - \varepsilon A_2 - \tfrac{1}{2}\varepsilon^2(-i)iA_1\\
&\quad = (1 - \tfrac{1}{2}\varepsilon^2)A_1 - \varepsilon A_2;
\end{aligned}
$$

similarly,

$$
\begin{aligned}
&(1 + i\varepsilon J_3 - \tfrac{1}{2}\varepsilon^2 J_3{}^2)A_2(1 - i\varepsilon J_3 - \tfrac{1}{2}\varepsilon^2 J_3{}^2)\\
&\quad = A_2 - i\varepsilon(A_2J_3 - J_3A_2) - \tfrac{1}{2}\varepsilon^2[(A_2J_3 - J_3A_2)J_3 - J_3(A_2J_3 - J_3A_2)]\\
&\quad = A_2 - i\varepsilon iA_1 - \tfrac{1}{2}\varepsilon^2 i(A_1J_3 - J_3A_1)\\
&\quad = A_2 + \varepsilon A_1 - \tfrac{1}{2}\varepsilon^2 i(-iA_2)\\
&\quad = (1 - \tfrac{1}{2}\varepsilon^2)A_2 + \varepsilon A_1
\end{aligned}
$$

and

$$
\begin{aligned}
&(1 + i\varepsilon J_3 - \tfrac{1}{2}\varepsilon^2 J_3{}^2)A_3(1 - i\varepsilon J_3 - \tfrac{1}{2}\varepsilon^2 J_3{}^2)\\
&\quad = (1 + i\varepsilon J_3 - \tfrac{1}{2}\varepsilon^2 J_3{}^2)(1 - i\varepsilon J_3 - \tfrac{1}{2}\varepsilon^2 J_3{}^2)A_3 = A_3
\end{aligned}
$$

in the approximation where only terms with ε and ε^2 are included.

We have seen some examples. For a spin and magnetic moment, the matrices J_1, J_2, J_3 are

$$\tfrac{1}{2}\Sigma_1, \tfrac{1}{2}\Sigma_2, \tfrac{1}{2}\Sigma_3.$$

They satisfy the characteristic commutation relations for matrices J_1, J_2, J_3 that correspond to rotations. The spin angular momentum, represented by the matrices

$$\tfrac{1}{2}\hbar\Sigma_1, \tfrac{1}{2}\hbar\Sigma_2, \tfrac{1}{2}\hbar\Sigma_3,$$

and the magnetic moment, represented by the matrices

$$\mu\Sigma_1, \mu\Sigma_2, \mu\Sigma_3,$$

are vector quantities. With each of these the matrices J_1, J_2, J_3 satisfy the commutation relations that characterize rotations of a vector quantity.

The orbital angular momentum provides another example. There the matrices J_1, J_2, J_3 are

$$\frac{1}{\hbar}L_1, \frac{1}{\hbar}L_2, \frac{1}{\hbar}L_3.$$

They satisfy the characteristic commutation relations for matrices J_1, J_2, J_3 that correspond to rotations. The position, represented by the matrices

$$Q_1, Q_2, Q_3;$$

the momentum, represented by the matrices

$$P_1, P_2, P_3;$$

and the orbital angular momentum, represented by the matrices

$$L_1, L_2, L_3$$

are vector quantities. With each of these, the matrices J_1, J_2, J_3 satisfy the commutation relations that characterize rotations of a vector quantity. This was demonstrated in Problem 20-2.

Some quantities are not changed by rotations. They are called scalar quantities. The magnitude, or the square of the magnitude of a vector quantity is a scalar quantity. For example, the square of the length of the position vector, represented by the matrix

$$R^2 = Q_1^2 + Q_2^2 + Q_3^2;$$

the square of the magnitude of the momentum, represented by the matrix

$$P_1^2 + P_2^2 + P_3^2;$$

and the square of the magnitude of the orbital angular momentum, represented by the matrix

$$L_1^2 + L_2^2 + L_3^2,$$

are scalar quantities. They are not changed by rotations.

The characteristic of a matrix that represents a scalar quantity is that it commutes with J_1, J_2, J_3. If B is a matrix that represents a scalar quantity, then for a rotation around the 3 axis, for example, we must have

$$\left(1 + i\varepsilon J_3 - \tfrac{1}{2}\varepsilon^2 J_3^{\,2}\right) B \left(1 - i\varepsilon J_3 - \tfrac{1}{2}\varepsilon^2 J_3^{\,2}\right) = B.$$

As before, when we work out the product on the left, keep the terms with ε, and ignore those with $\varepsilon^2, \varepsilon^3, \ldots,$ we get

$$B - i\varepsilon(BJ_3 - J_3 B) = B.$$

Again, this is not an approximation. It has to hold when higher powers of ε are included if the full equation is to hold for any choice of the small real number ε. It follows that B commutes with J_3. From rotations around the 1 and 2 axes, it follows similarly that if B represents a scalar quantity, then B commutes with J_1 and J_2. Conversely, if B is a matrix that commutes with J_1, J_2, J_3, then it is not changed by rotations, so it represents a scalar quantity; for a rotation around the 3 axis, for example, we get

$$\begin{aligned}&\left(1 + i\varepsilon J_3 - \tfrac{1}{2}\varepsilon^2 J_3^{\,2}\right) B \left(1 - i\varepsilon J_3 - \tfrac{1}{2}\varepsilon^2 J_3^{\,2}\right)\\ &= \left(1 + i\varepsilon J_3 - \tfrac{1}{2}\varepsilon^2 J_3^{\,2}\right)\left(1 - i\varepsilon J_3 - \tfrac{1}{2}\varepsilon^2 J_3^{\,2}\right) B = B.\end{aligned}$$

In each example we have considered, the matrix representing the scalar quantity commutes with the matrices L_1, L_2, L_3 for orbital angular momentum and therefore commutes with the matrices J_1, J_2, J_3 for that example. We showed this for the square of the magnitude of the angular momentum. For the square of the length of the position vector and the square of the magnitude of the momentum, it is demonstrated in Problem 20-3.

Thus we see there are two properties of angular momentum that correspond to rotations. The matrices that represent angular momentum divided by $\hbar$ satisfy the characteristic commutation relations for matrices J_1, J_2, J_3 that correspond to rotations. They also satisfy equations that characterize rotations of the physical quantities. They commute with each matrix that represents a scalar quantity. With matrices that represent a vector quantity, they satisfy the commutation relations that characterize rotations of a vector quantity. In particular, we see the two properties together in the example of orbital angular momentum with position and momentum.

PROBLEMS

24-1. Show that multiplying the matrix $1 - i\varepsilon M_3 - \frac{1}{2}\varepsilon^2 M_3^{\,2}$ by the similar matrix with ε replaced by ε' gives another matrix of this form where

ε is replaced by $\varepsilon + \varepsilon'$, in the approximation where terms quadratic in ε and ε' are included but terms with higher powers are ignored. This shows that the product of two rotations measured by ε and ε' is the rotation measured by $\varepsilon + \varepsilon'$. In the particular case where ε' is $-\varepsilon$, the product is the identity rotation, so the second rotation is the inverse of the first.

24-2. Do the same as in the last problem for the matrix

$$1 - i\varepsilon J_3 - \tfrac{1}{2}\varepsilon^2 J_3^{\,2}.$$

24-3. Let A_1, A_2, A_3 be matrices that represent projections of a vector quantity, so A_1, A_2, A_3 and J_1, J_2, J_3 satisfy the commutation relations that characterize rotations of a vector quantity. Write out the commutation relations between A_1, A_2, A_3 and J_1, J_2 that characterize rotations of the vector quantity around the 1 and 2 axes. Show that

$$A_1^{\,2} + A_2^{\,2} + A_3^{\,2}$$

commutes with J_1, J_2, J_3. This shows that the square of the magnitude of any vector quantity is a scalar quantity.

24-4. Suppose A_1, A_2, A_3 and K_1, K_2, K_3 are matrices that represent projections of two vector quantities and B is a matrix that represents a scalar quantity, so B commutes with J_1, J_2, J_3 and each set of matrices A_1, A_2, A_3 and K_1, K_2, K_3 satisfies the commutation relations with J_1, J_2, J_3 that characterize rotations of a vector quantity. Show that each set of matrices

$$A_1 + K_1, \quad A_2 + K_2, \quad A_3 + K_3$$

and

$$BA_1, BA_2, BA_3$$

satisfies the commutation relations with J_1, J_2, J_3 that characterize rotations of a vector quantity. Show that the matrix

$$A_1K_1 + A_2K_2 + A_3K_3$$

commutes with J_1, J_2, J_3.

24-5. Let L_1, L_2, L_3 be the matrices that represent orbital angular momentum, and let

$$J_1' = -\frac{1}{\hbar}L_2, \quad J_2' = \frac{1}{\hbar}L_1, \quad J_3' = \frac{1}{\hbar}L_3.$$

Show that J_1', J_2', J_3' satisfy the commutation relations characteristic of matrices J_1, J_2, J_3 that correspond to rotations, but with the matrices Q_1, Q_2, Q_3 and P_1, P_2, P_3 that represent position and momentum, J_1', J_2', J_3' do not satisfy the commutation relations that characterize rotations of vector quantities.

24-6. To describe the spin and magnetic moment of a particle such as an electron or proton together with the motion in three-dimensional space, we need matrices $\Sigma_1, \Sigma_2, \Sigma_3$ as well as matrices Q_1, Q_2, Q_3 and P_1, P_2, P_3 that represent position and momentum. The matrices $\Sigma_1, \Sigma_2, \Sigma_3$ commute with Q_1, Q_2, Q_3 and P_1, P_2, P_3, and the multiplication rules for $\Sigma_1, \Sigma_2, \Sigma_3$ and the commutation relations for Q_1, Q_2, Q_3 and P_1, P_2, P_3 are the same as before. The total angular momentum is the sum of the orbital angular momentum and spin angular momentum. Then

$$J_1 = \frac{1}{\hbar}L_1 + \tfrac{1}{2}\Sigma_1,$$

$$J_2 = \frac{1}{\hbar}L_2 + \tfrac{1}{2}\Sigma_2,$$

$$J_3 = \frac{1}{\hbar}L_3 + \tfrac{1}{2}\Sigma_3.$$

Show that these matrices satisfy the commutation relations characteristic of matrices J_1, J_2, J_3 that correspond to rotations. The spin angular momentum represented by the matrices

$$\tfrac{1}{2}\hbar\Sigma_1, \tfrac{1}{2}\hbar\Sigma_2, \tfrac{1}{2}\hbar\Sigma_3;$$

the magnetic moment represented by

$$\mu\Sigma_1, \mu\Sigma_2, \mu\Sigma_3;$$

the position represented by Q_1, Q_2, Q_3; the momentum represented by P_1, P_2, P_3; and the orbital angular momentum represented by

L_1, L_2, L_3 are vector quantities. Show that with each of these the matrices J_1, J_2, J_3 satisfy the commutation relations that characterize rotations of a vector quantity.

24-7. Let A_1, A_2, A_3 be matrices that represent a vector quantity, so with J_1, J_2, J_3 they satisfy the commutation relations that characterize rotations of a vector quantity. For example, A_1, A_2, A_3 could be matrices Q_1, Q_2, Q_3 representing position coordinates. Let

$$U = \left(1 + i\varepsilon J_2 - \tfrac{1}{2}\varepsilon^2 J_2{}^2\right)\left(1 + i\varepsilon J_1 - \tfrac{1}{2}\varepsilon^2 J_1{}^2\right)$$

$$\times\left(1 - i\varepsilon J_2 - \tfrac{1}{2}\varepsilon^2 J_2{}^2\right)\left(1 - i\varepsilon J_1 - \tfrac{1}{2}\varepsilon^2 J_1{}^2\right),$$

$$N = \left(1 + i\varepsilon M_2 - \tfrac{1}{2}\varepsilon^2 M_2{}^2\right)\left(1 + i\varepsilon M_1 - \tfrac{1}{2}\varepsilon^2 M_1{}^2\right)$$

$$\times\left(1 - i\varepsilon M_2 - \tfrac{1}{2}\varepsilon^2 M_2{}^2\right)\left(1 - i\varepsilon M_1 - \tfrac{1}{2}\varepsilon^2 M_1{}^2\right)$$

and

$$X = \begin{pmatrix} x_1 \\ x_2 \\ x_3 \end{pmatrix}.$$

Consider the single-column matrix NX obtained multiplying the single-column matrix X with the 3×3 matrix N. Show that the formulas for $U^{-1}A_1U$, $U^{-1}A_2U$, $U^{-1}A_3U$ in terms of A_1, A_2, A_3 and ε are the same as the formulas for the three elements of NX in terms of x_1, x_2, x_3 and ε. You do not need to write out the complete formulas to see they are the same. This shows why the order of the factors in U and N is the same rather than opposite, even though they are used so differently that the factors enter U in succession from left to right and N from right to left.

REFERENCES

1. B. L. van der Waerden, in *Theoretical Physics in the Twentieth Century: A Memorial Volume to Wolfgang Pauli*, edited by M. Fierz and V. F. Weisskopf. Interscience, New York, 1960, pp. 221–228.
2. E. P. Wigner, in private conversation, 24 April 1984.
3. V. Bargmann, *Ann. Math.* **59**, 1 (1954).
4. V. Bargmann, *J. Math. Phys.* **5**, 862 (1964).

5. M. Hamermesh, *Group Theory and Its Applications to Physical Problems*. Addison-Wesley, Reading, Massachusetts, 1962, Chap. 12.
6. J. M. Jauch, *Foundations of Quantum Mechanics*. Addison-Wesley, Reading, Massachusetts, 1968, Chaps. 9–10, 12–14.
7. T. F. Jordan, *Linear Operators for Quantum Mechanics*. Wiley, New York, 1969, Chap. 7.
8. E. P. Wigner, *Ann. Math.* **40**, 149 (1939).
9. E. P. Wigner, *Group Theory*. Academic Press, New York, 1959, particularly the Appendix to Chap. 20.

25 CHANGES IN SPACE LOCATION

The same ideas that we developed to discuss rotations can be used for other changes of space and time coordinates. Here we consider changes of location in space. That will lead to the commutation relations between position and momentum matrices and show us the origin of Born's "strange equation."

Let x be a position coordinate of an object, measured along some line from some reference point.

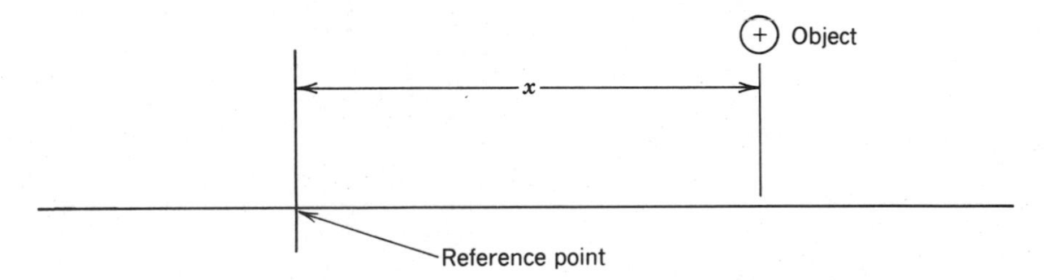

If the object were a distance ε farther along the direction of that line, the coordinate would be changed to $x + \varepsilon$.

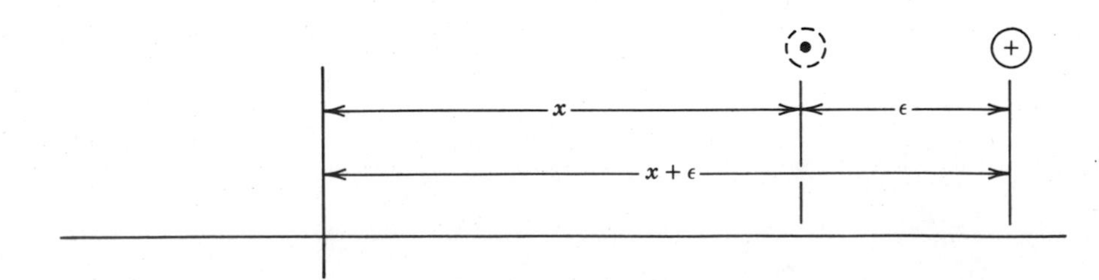

The same change of coordinate results from moving the reference point a distance ε in the opposite direction.

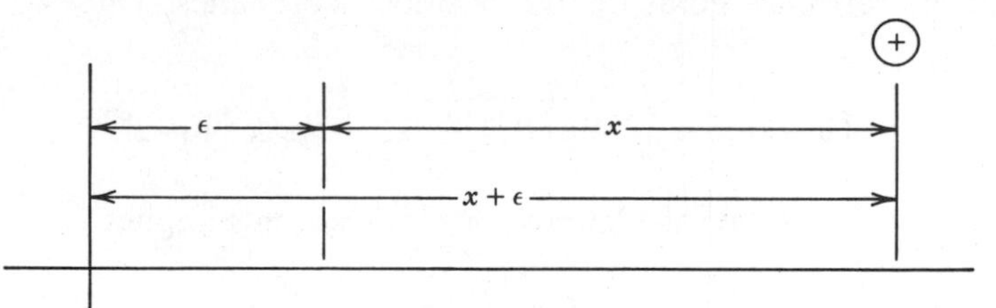

It relates descriptions of the same object by two observers who use different reference points. You can look at it either way. The change of coordinate is called a space translation. We consider the change in the description of a system in quantum mechanics made by this change of coordinate, assuming nothing else is changed. We have to accommodate such changes because the choice of coordinates is often rather arbitrary; one reference point may be no better than another.

In quantum mechanics the position coordinate is represented by a matrix Q. We consider it being changed to $Q + \varepsilon$. For distances ε that are small in some suitable units, this is generally done by multiplying with a matrix

$$1 - i\varepsilon K - \tfrac{1}{2}\varepsilon^2 K^2$$

and its inverse

$$1 + i\varepsilon K - \tfrac{1}{2}\varepsilon^2 K^2,$$

changing Q to

$$\begin{aligned}(1 + i\varepsilon K - \tfrac{1}{2}\varepsilon^2 K^2)Q(1 - i\varepsilon K - \tfrac{1}{2}\varepsilon^2 K^2)\\ = Q - i\varepsilon(QK - KQ) - \tfrac{1}{2}\varepsilon^2[(QK - KQ)K - K(QK - KQ)].\end{aligned}$$

The right side is obtained by multiplying out the product on the left. We keep terms with ε and ε^2 and ignore those with $\varepsilon^3, \varepsilon^4, \ldots$, just as we did for small rotations. For the result to be $Q + \varepsilon$ for any choice of the small number ε, we must have

$$QK - KQ = i.$$

Then the term with ε^2 is zero and we do get

$$(1 + i\varepsilon K - \tfrac{1}{2}\varepsilon^2 K^2)Q(1 - i\varepsilon K - \tfrac{1}{2}\varepsilon^2 K^2) = Q + \varepsilon.$$

We assume the matrix P representing momentum is not changed. The mass and velocity of an object are the same in two descriptions that differ only in the reference point for the position coordinates. Thus we assume that

$$\left(1 + i\varepsilon K - \tfrac{1}{2}\varepsilon^2 K^2\right) P \left(1 - i\varepsilon K - \tfrac{1}{2}\varepsilon^2 K^2\right) = P.$$

By comparing with the calculations for Q, we see this implies

$$PK - KP = 0.$$

If the position and momentum are represented by matrices Q and P that satisfy the commutation relation

$$QP - PQ = i\hbar,$$

then the changes of these matrices corresponding to the changes of the position coordinate are obtained by letting

$$K = \frac{1}{\hbar} P.$$

This is the meaning of Born's "strange equation." The matrix P plays two roles. It represents the momentum of the object in the direction the coordinate is measured. It also provides the matrix K that we use to make the changes in the matrices representing position and momentum corresponding to changes in the position coordinate. That is where the commutation relations come in.

Now we extend this to three-dimensional position and momentum. Just as for one dimension, this leads to the commutation relations between position and momentum matrices. In addition, we find that the commuting of matrices representing different projections of the momentum corresponds to the commuting of space translations in different directions, and the commutation relations between matrices representing momentum and angular momentum correspond to relations between space translations and rotations.

Consider a change of the position coordinate in the 1 direction. This is also called a translation in the 1 direction. The matrix Q_1 representing the position coordinate in the 1 direction is changed to $Q_1 + \varepsilon$. The matrices Q_2 and Q_3 representing the position coordinates in the 2 and 3 directions are not changed. The matrices P_1, P_2, P_3 representing momentum are not changed. For small distances ε, this change is made with a matrix K_1 that gives

$$(1 + i\varepsilon K_1 - \tfrac{1}{2}\varepsilon^2 K_1{}^2)Q_1(1 - i\varepsilon K_1 - \tfrac{1}{2}\varepsilon^2 K_1{}^2) = Q_1 + \varepsilon,$$

$$(1 + i\varepsilon K_1 - \tfrac{1}{2}\varepsilon^2 K_1{}^2)Q_2(1 - i\varepsilon K_1 - \tfrac{1}{2}\varepsilon^2 K_1{}^2) = Q_2,$$

$$(1 + i\varepsilon K_1 - \tfrac{1}{2}\varepsilon^2 K_1{}^2)Q_3(1 - i\varepsilon K_1 - \tfrac{1}{2}\varepsilon^2 K_1{}^2) = Q_3,$$

$$(1 + i\varepsilon K_1 - \tfrac{1}{2}\varepsilon^2 K_1{}^2)P_1(1 - i\varepsilon K_1 - \tfrac{1}{2}\varepsilon^2 K_1{}^2) = P_1,$$

$$(1 + i\varepsilon K_1 - \tfrac{1}{2}\varepsilon^2 K_1{}^2)P_2(1 - i\varepsilon K_1 - \tfrac{1}{2}\varepsilon^2 K_1{}^2) = P_2,$$

$$(1 + i\varepsilon K_1 - \tfrac{1}{2}\varepsilon^2 K_1{}^2)P_3(1 - i\varepsilon K_1 - \tfrac{1}{2}\varepsilon^2 K_1{}^2) = P_3.$$

By comparing with the calculations for one dimension, we see this means

$$Q_1 K_1 - K_1 Q_1 = i,$$

$$Q_2 K_1 - K_1 Q_2 = 0,$$

$$Q_3 K_1 - K_1 Q_3 = 0,$$

$$P_1 K_1 - K_1 P_1 = 0,$$

$$P_2 K_1 - K_1 P_2 = 0,$$

$$P_3 K_1 - K_1 P_3 = 0.$$

Similarly, the changes of matrices corresponding to small changes of the position coordinate in the 2 direction are made with a matrix K_2 that commutes with Q_1, Q_3, P_1, P_2, P_3 and gives

$$Q_2 K_2 - K_2 Q_2 = i,$$

and the changes of matrices corresponding to small changes of the position coordinate in the 3 direction are made with a matrix K_3 that commutes with Q_1, Q_2, P_1, P_2, P_3 and gives

$$Q_3 K_3 - K_3 Q_3 = i.$$

This is the general scheme for representing space translations in quantum mechanics. It follows from assumptions that are valid for any isolated system. They are analogous to the assumptions outlined for rotations in the last chapter. These assumptions are discussed again in Chapter 28 for rotations, translations, and changes in time and velocity, all together.

Translations in the 1, 2, 3 directions are represented by the matrices

$$1 - i\varepsilon K_1 - \tfrac{1}{2}\varepsilon^2 K_1{}^2,$$

$$1 - i\varepsilon K_2 - \tfrac{1}{2}\varepsilon^2 K_2{}^2,$$

$$1 - i\varepsilon K_3 - \tfrac{1}{2}\varepsilon^2 K_3{}^2$$

the same way rotations around the 1, 2, 3 axes are represented by the matrices

$$1 - i\varepsilon J_1 - \tfrac{1}{2}\varepsilon^2 J_1{}^2,$$

$$1 - i\varepsilon J_2 - \tfrac{1}{2}\varepsilon^2 J_2{}^2,$$

$$1 - i\varepsilon J_3 - \tfrac{1}{2}\varepsilon^2 J_3{}^2.$$

Products of these translations are represented by the corresponding products of the matrices, just as products of the rotations are represented by products of the matrices that represent them. The product of two translations is the change of position coordinates made by doing first one translation and then the other.

Each translation has an inverse. It is the translation the same distance in the opposite direction. If the object, or the reference point, is moved a distance ε in one direction and then the same distance in the opposite direction, the result is no change at all. No change at all is called the identity translation, so the product of a translation and its inverse is the identity translation.

Translations in different directions commute. Doing first a translation in the 1 direction and then a translation in the 2 direction gives the same result as doing first the translation in the 2 direction and then the translation in the 1 direction.

Likewise, the matrices that represent translations commute. For example,

$$\left(1 - i\varepsilon K_2 - \tfrac{1}{2}\varepsilon^2 K_2{}^2\right)\left(1 - i\varepsilon K_1 - \tfrac{1}{2}\varepsilon^2 K_1{}^2\right)$$

$$= \left(1 - i\varepsilon K_1 - \tfrac{1}{2}\varepsilon^2 K_1{}^2\right)\left(1 - i\varepsilon K_2 - \tfrac{1}{2}\varepsilon^2 K_2{}^2\right).$$

This follows because each product of the matrices corresponds to the product of the translations; both matrix products are the matrix that represents the product of the translations. This is part of the general scheme that follows from assumptions valid for any isolated system.

This implies the matrices K_1, K_2, K_3 commute. If we work out the products in the last equation, keep only the terms with ε and ε^2, and cancel the terms that are the same on both sides, we get

$$-\varepsilon^2 K_2 K_1 = -\varepsilon^2 K_1 K_2.$$

This is not an approximation. It has to hold when higher powers of ε are included because the matrices representing translations commute for any choice of the small real number ε. Therefore

$$K_2 K_1 = K_1 K_2.$$

It follows similarly that all the matrices K_1, K_2, K_3 commute with each other.

Another way to see this is to consider the matrix product

$$\begin{aligned}&\left(1 + i\varepsilon K_2 - \tfrac{1}{2}\varepsilon^2 K_2{}^2\right)\left(1 + i\varepsilon K_1 - \tfrac{1}{2}\varepsilon^2 K_1{}^2\right)\\ &\times\left(1 - i\varepsilon K_2 - \tfrac{1}{2}\varepsilon^2 K_2{}^2\right)\left(1 - i\varepsilon K_1 - \tfrac{1}{2}\varepsilon^2 K_1{}^2\right)\\ &\qquad = 1 + \varepsilon^2 (K_1 K_2 - K_2 K_1).\end{aligned}$$

The right side is obtained by working out the product on the left the same as we did for rotations. Now the four matrices in the product represent four space translations. It is easy to find the product of these translations. Since they commute, we can multiply them in any order. One is a translation a distance ε in the 1 direction. Another is a translation a distance $-\varepsilon$ in the 1 direction. These are inverses of each other, so their product is the identity translation. The other two are translations by ε and $-\varepsilon$ in the 2 direction, which are inverses of each other, so their product also is the identity translation. Therefore the product of the four translations is the identity translation. It is represented by the matrix 1. Then

$$1 + \varepsilon^2 (K_1 K_2 - K_2 K_1) = 1$$

because the product of the four matrices representing the four translations is the matrix that represents the product of the four translations. Again, this is part of the general scheme that follows from assumptions valid for any isolated system. The last equation is not an approximation. It has to hold when higher powers of ε are included because the full equation holds for any choice of the small real number ε. Therefore

$$K_1K_2 - K_2K_1 = 0.$$

The fact that translations in different directions commute is reflected by K_1, K_2, K_3 commuting with each other.

We combine translations and rotations by considering both as changes of coordinates. The product of a translation and a rotation is the change of coordinates made by doing first one and then the other. Products of the products are defined the same way. These products of translations and rotations are represented by the corresponding products of the matrices that represent them.

A translation in the 3 direction commutes with a rotation around the 3 axis because the translation changes only x_3 and the rotation changes only x_1 and x_2. Likewise, the matrices that represent them commute, which means

$$J_3K_3 = K_3J_3$$

or

$$J_3K_3 - K_3J_3 = 0.$$

This is obtained the same way as that K_1, K_2, K_3 commute with each other, from the general scheme that follows from assumptions valid for any isolated system, extended here to include translations and rotations together. It follows similarly that

$$J_1K_1 - K_1J_1 = 0$$

and

$$J_2K_2 - K_2J_2 = 0.$$

That these matrices commute is a reflection of the fact that a translation and a rotation commute if their axes are the same.

We also obtain commutation relations for matrices such as J_3 and K_2 that do not commute. We consider the matrix product

$$\begin{aligned}&\left(1 - i\varepsilon K_2 - \tfrac{1}{2}\varepsilon^2K_2^2\right)\left(1 - i\varepsilon J_3 - \tfrac{1}{2}\varepsilon^2J_3^2\right)\\ &\times\left(1 + i\varepsilon K_2 - \tfrac{1}{2}\varepsilon^2K_2^2\right)\left(1 + i\varepsilon J_3 - \tfrac{1}{2}\varepsilon^2J_3^2\right) = 1 + \varepsilon^2\left(J_3K_2 - K_2J_3\right).\end{aligned}$$

Again, the right side is obtained by working out the product on the left the same as we did for rotations. The four matrices in the product represent a translation in the 2 direction, a rotation around the 3 axis, and their inverses. None of these translations and rotations changes x_3. We find their product by computing the changes of x_1 and x_2 produced by the four translations and rotations in succession. We work from left to right because that is the way matrices enter the product UV in the succession of changes of matrices representing physical quantities from B to $U^{-1}BU$ to $V^{-1}U^{-1}BUV$.

The product of the four translations and rotations is not the identity because they do not commute. First there is a translation in the 2 direction, which changes x_1 and x_2 to

$$x_1 \quad \text{and} \quad x_2 + \varepsilon.$$

Then there is a rotation around the 3 axis, which changes x_1 and x_2 in these to give

$$\left(1 - \tfrac{1}{2}\varepsilon^2\right)x_1 - \varepsilon x_2 \quad \text{and} \quad \varepsilon x_1 + \left(1 - \tfrac{1}{2}\varepsilon^2\right)x_2 + \varepsilon.$$

We are keeping only terms with ε and ε^2. Next there is the inverse translation in the 2 direction, which changes x_1 and x_2 to x_1 and $x_2 - \varepsilon$ and gives

$$\left(1 - \tfrac{1}{2}\varepsilon^2\right)x_1 - \varepsilon(x_2 - \varepsilon) \quad \text{and} \quad \varepsilon x_1 + \left(1 - \tfrac{1}{2}\varepsilon^2\right)(x_2 - \varepsilon) + \varepsilon.$$

Finally there is the inverse rotation around the 3 axis, which changes x_1 and x_2 in these to give

$$\left(1 - \tfrac{1}{2}\varepsilon^2\right)\left[\left(1 - \tfrac{1}{2}\varepsilon^2\right)x_1 + \varepsilon x_2\right] - \varepsilon\left[-\varepsilon x_1 + \left(1 - \tfrac{1}{2}\varepsilon^2\right)x_2 - \varepsilon\right]$$

$$= x_1 - 2\tfrac{1}{2}\varepsilon^2 x_1 + \varepsilon x_2 + \varepsilon^2 x_1 - \varepsilon x_2 + \varepsilon^2 = x_1 + \varepsilon^2$$

and

$$\varepsilon\left[\left(1 - \tfrac{1}{2}\varepsilon^2\right)x_1 + \varepsilon x_2\right] + \left(1 - \tfrac{1}{2}\varepsilon^2\right)\left[-\varepsilon x_1 + \left(1 - \tfrac{1}{2}\varepsilon^2\right)x_2 - \varepsilon\right] + \varepsilon$$

$$= \varepsilon x_1 + \varepsilon^2 x_2 - \varepsilon x_1 + x_2 - 2\tfrac{1}{2}\varepsilon^2 x_2 - \varepsilon + \varepsilon = x_2.$$

This is the result of the four translations and rotations. Altogether, x_1 is changed to $x_1 + \varepsilon^2$ and x_2 is not changed. Since x_3 also is not changed, the product of the four translations and rotations is just a translation a distance

ε^2 in the 1 direction. It is represented by the matrix

$$1 - i\varepsilon^2 K_1.$$

Then

$$1 + \varepsilon^2(J_3K_2 - K_2J_3) = 1 - i\varepsilon^2 K_1$$

because the product of the four matrices is the matrix that represents the product of the four translations and rotations. This is part of the general scheme that follows from assumptions valid for any isolated system, extended here to include translations and rotations together. The last equation is not an approximation. It has to hold when higher powers of ε are included because the full equation holds for any choice of the small real number ε. Therefore

$$J_3K_2 - K_2J_3 = -iK_1.$$

It follows similarly that

$$J_1K_2 - K_2J_1 = iK_3, \qquad J_2K_1 - K_1J_2 = -iK_3,$$

$$J_2K_3 - K_3J_2 = iK_1,$$

$$J_3K_1 - K_1J_3 = iK_2, \qquad J_1K_3 - K_3J_1 = -iK_2.$$

Thus we see that these commutation relations correspond to multiplications of translations and rotations.

The way we made changes of coordinates in succession to find the product of the four translations and rotations is the same way position matrices Q_1, Q_2, Q_3 are changed by multiplication in succession with the matrices that represent the translations and rotations. Since Q_1, Q_2, Q_3 represent a vector quantity, we assume they satisfy the commutation relations with J_1, J_2, J_3 that characterize rotations of a vector quantity. From these and the commutation relations of Q_1, Q_2, Q_3 with K_1, K_2, K_3, we find that in the first step, the space translation in the 2 direction, Q_1 and Q_2 are changed to

$$\left(1 + i\varepsilon K_2 - \tfrac{1}{2}\varepsilon^2 K_2{}^2\right)Q_1\left(1 - i\varepsilon K_2 - \tfrac{1}{2}\varepsilon^2 K_2{}^2\right) = Q_1$$

and

$$\left(1 + i\varepsilon K_2 - \tfrac{1}{2}\varepsilon^2 K_2{}^2\right)Q_2\left(1 - i\varepsilon K_2 - \tfrac{1}{2}\varepsilon^2 K_2{}^2\right) = Q_2 + \varepsilon$$

and then in the second step, the rotation around the 3 axis, these are changed to

$$\begin{aligned}
&\left(1 + i\varepsilon J_3 - \tfrac{1}{2}\varepsilon^2 J_3^{\,2}\right)\left(1 + i\varepsilon K_2 - \tfrac{1}{2}\varepsilon^2 K_2^{\,2}\right)\\
&\quad\times Q_1\left(1 - i\varepsilon K_2 - \tfrac{1}{2}\varepsilon^2 K_2^{\,2}\right)\left(1 - i\varepsilon J_3 - \tfrac{1}{2}\varepsilon^2 J_3^{\,2}\right)\\
&\quad= \left(1 + i\varepsilon J_3 - \tfrac{1}{2}\varepsilon^2 J_3^{\,2}\right)Q_1\left(1 - i\varepsilon J_3 - \tfrac{1}{2}\varepsilon^2 J_3^{\,2}\right) = \left(1 - \tfrac{1}{2}\varepsilon^2\right)Q_1 - \varepsilon Q_2
\end{aligned}$$

and

$$\begin{aligned}
&\left(1 + i\varepsilon J_3 - \tfrac{1}{2}\varepsilon^2 J_3^{\,2}\right)\left(1 + i\varepsilon K_2 - \tfrac{1}{2}\varepsilon^2 K_2^{\,2}\right)\\
&\quad\times Q_2\left(1 - i\varepsilon K_2 - \tfrac{1}{2}\varepsilon^2 K_2^{\,2}\right)\left(1 - i\varepsilon J_3 - \tfrac{1}{2}\varepsilon^2 J_3^{\,2}\right)\\
&\quad= \left(1 + i\varepsilon J_3 - \tfrac{1}{2}\varepsilon^2 J_3^{\,2}\right)\left(Q_2 + \varepsilon\right)\left(1 - i\varepsilon J_3 - \tfrac{1}{2}\varepsilon^2 J_3^{\,2}\right)\\
&\quad= \varepsilon Q_1 + \left(1 - \tfrac{1}{2}\varepsilon^2\right)Q_2 + \varepsilon.
\end{aligned}$$

This is the same way x_1 and x_2 are changed. We can see similarly that Q_1 and Q_2 are changed the same way as x_1 and x_2 in the third and fourth steps. This confirms that we are calculating the different kinds of products in a consistent way.

We have found three different origins of commutation relations for the matrices K_1, K_2, K_3. The commutation relations of K_1, K_2, K_3 with Q_1, Q_2, Q_3 and P_1, P_2, P_3 correspond to the way position and momentum are changed or not changed by space translations. The commuting of K_1, K_2, K_3 with each other corresponds to the fact that translations in different directions commute. The commutation relations of K_1, K_2, K_3 with J_1, J_2, J_3 correspond to relations between space translations and rotations.

All these commutation relations are satisfied if

$$K_1 = \frac{1}{\hbar}P_1, \qquad K_2 = \frac{1}{\hbar}P_2, \qquad K_3 = \frac{1}{\hbar}P_3$$

and

$$J_1 = \frac{1}{\hbar}L_1, \qquad J_2 = \frac{1}{\hbar}L_2, \qquad J_3 = \frac{1}{\hbar}L_3,$$

where L_1, L_2, L_3 are the matrices representing orbital angular momentum made from Q_1, Q_2, Q_3 and P_1, P_2, P_3. Then all the commutation relations follow from those of Q_1, Q_2, Q_3 and P_1, P_2, P_3. This shows us more fully the meaning of these "strange equations."

Consider the equation

$$L_3P_2 - P_2L_3 = -i\hbar P_1.$$

It relates matrices that represent momentum and angular momentum. It might seem rather mysterious if that were all we knew about it, but we know more. After dividing by $\hbar$, we can write the same equation as

$$J_3P_2 - P_2J_3 = -iP_1.$$

It tells us how a matrix representing momentum is changed by a rotation. We can also write this equation as

$$L_3K_2 - K_2L_3 = -iP_1.$$

It tells us how a matrix representing angular momentum is changed by a translation. This is to be developed in Problem 25-1. After dividing by $\hbar$ again, we can write this same equation as

$$J_3K_2 - K_2J_3 = -iK_1.$$

It corresponds to multiplications of translations and rotations. This is an example of a wonderful economy in quantum mechanics. The same thing is used several different ways.

PROBLEMS

25-1. Let

$$U = 1 - i\varepsilon K_2 - \tfrac{1}{2}\varepsilon^2 K_2{}^2.$$

Calculate $U^{-1}L_3U$. Do this two different ways. First use

$$L_3 = Q_1P_2 - Q_2P_1,$$

the fact that K_2 and U commute with P_1 and P_2, and the formulas for $U^{-1}Q_1U$ and $U^{-1}Q_2U$. Then use

$$U^{-1}L_3U = L_3 - i\varepsilon(L_3K_2 - K_2L_3)$$
$$- \tfrac{1}{2}\varepsilon^2[(L_3K_2 - K_2L_3)K_2 - K_2(L_3K_2 - K_2L_3)]$$

and the commutation relation

$$L_3K_2 - K_2L_3 = -iP_1.$$

If you get the same answer, it shows the commutation relation does correspond to the way angular momentum is changed by a translation.

25-2. Consider a particle with a spin and magnetic moment described by matrices $\Sigma_1, \Sigma_2, \Sigma_3$ in addition to position and momentum represented by matrices Q_1, Q_2, Q_3 and P_1, P_2, P_3, as in Problem 24-6. Then J_1, J_2, J_3 are the same as there. Let

$$K_1 = \frac{1}{\hbar}P_1, \qquad K_2 = \frac{1}{\hbar}P_2, \qquad K_3 = \frac{1}{\hbar}P_3.$$

Show that these matrices K_1, K_2, K_3 satisfy all the commutation relations we have considered: with Q_1, Q_2, Q_3 and P_1, P_2, P_3; with each other; and with J_1, J_2, J_3. In addition, K_1, K_2, K_3 commute with $\Sigma_1, \Sigma_2, \Sigma_3$. This corresponds to the fact that the spin and magnetic moment are not changed by space translations.

26 CHANGES IN TIME

Now we consider how the description of a system changes from one time to another. It is just like the change made by a translation from one space location to another or a rotation from one orientation to another. There is a matrix Ω that is used for changes in time the same way K_1, K_2, K_3 are used for space translations and J_1, J_2, J_3 for rotations. When the time changes by an amount ε that is small in some suitable units, each matrix B that represents a physical quantity at the initial time is changed to a corresponding matrix

$$\left(1 + i\varepsilon\Omega - \tfrac{1}{2}\varepsilon^2\Omega^2\right)B\left(1 - i\varepsilon\Omega - \tfrac{1}{2}\varepsilon^2\Omega^2\right)$$

that represents the quantity at time ε later. As usual, we keep only terms with ε and ε^2.

This also describes a change of the time coordinate. It relates descriptions of the same system by two observers who set their clocks differently. If my clock lags behind yours by ε, changing from your description of the system at noon to my description at the time I call noon is the same as changing between your descriptions at noon and noon + ε.

Setting the clock is like choosing a reference point for the space coordinates.

We combine changes in time with space translations and rotations by considering all of them as changes of space and time coordinates. The product of a change in time and a space translation or rotation is the change of space and time coordinates made by doing first one and then the other.

Changes in time commute with space translations and rotations because changes in time change only the time coordinate and space translations and rotations change only the space coordinates. Likewise, the corresponding matrices commute. The matrix Ω commutes with K_1, K_2, K_3 and J_1, J_2, J_3.

This is obtained the same as that K_1, K_2, K_3 commute with each other, as part of the general scheme, which is extended here to include space translations, rotations, and changes in time together.

If B commutes with Ω, then B represents a quantity that does not change in time, because

$$\left(1 + i\varepsilon\Omega - \tfrac{1}{2}\varepsilon^2\Omega^2\right)B\left(1 - i\varepsilon\Omega - \tfrac{1}{2}\varepsilon^2\Omega^2\right)$$

$$= B\left(1 + i\varepsilon\Omega - \tfrac{1}{2}\varepsilon^2\Omega^2\right)\left(1 - i\varepsilon\Omega - \tfrac{1}{2}\varepsilon^2\Omega^2\right) = B.$$

The quantity is constant in time.

Since K_1, K_2, K_3 and J_1, J_2, J_3 commute with Ω, they represent quantities that are constant in time. They are typically the momentum and angular momentum divided by $\hbar$. Looking at it the other way, we see that Ω commutes with K_1, K_2, K_3 and J_1, J_2, J_3, which means Ω represents a quantity that is not changed by space translations and rotations. All this reflects the fact that changes of the time coordinate commute with space translations and rotations.

Since Ω commutes with Ω, it represents a quantity that is constant in time. Typically, Ω is $(1/\hbar)H$, where H is the matrix that represents the energy.

This is the general scheme. It follows from assumptions that are valid for any isolated system. Again, they are analogous to the assumptions outlined in Chapter 24 for rotations. They will be discussed in Chapter 28. Now we consider some particular cases.

Suppose the system is a single isolated particle that moves with constant velocity. Let m be the mass of the particle and let the momentum at some time be represented by matrices P_1, P_2, P_3. Then, since the momentum is the velocity multiplied by the mass, the velocity is represented by the matrices

$$V_1 = \frac{1}{m}P_1, \qquad V_2 = \frac{1}{m}P_2, \qquad V_3 = \frac{1}{m}P_3.$$

Since the velocity is constant, the momentum is constant and the matrices P_1, P_2, P_3 are not changed in time. Therefore

$$\left(1 + i\varepsilon\Omega - \tfrac{1}{2}\varepsilon^2\Omega^2\right)P_1\left(1 - i\varepsilon\Omega - \tfrac{1}{2}\varepsilon^2\Omega^2\right) = P_1,$$

$$\left(1 + i\varepsilon\Omega - \tfrac{1}{2}\varepsilon^2\Omega^2\right)P_2\left(1 - i\varepsilon\Omega - \tfrac{1}{2}\varepsilon^2\Omega^2\right) = P_2,$$

$$\left(1 + i\varepsilon\Omega - \tfrac{1}{2}\varepsilon^2\Omega^2\right)P_3\left(1 - i\varepsilon\Omega - \tfrac{1}{2}\varepsilon^2\Omega^2\right) = P_3.$$

Let the position coordinates at some time be represented by matrices

Q_1, Q_2, Q_3. The distance the particle moves in a given time is the velocity multiplied by the time. During a small time interval of duration ε, a position coordinate changes by ε times the projection of the velocity in the direction the coordinate is measured. Then, for example, Q_1 changes to

$$Q_1 + \varepsilon V_1 = Q_1 + \varepsilon \frac{1}{m} P_1,$$

so

$$\left(1 + i\varepsilon\Omega - \tfrac{1}{2}\varepsilon^2\Omega^2\right) Q_1 \left(1 - i\varepsilon\Omega - \tfrac{1}{2}\varepsilon^2\Omega^2\right) = Q_1 + \varepsilon \frac{1}{m} P_1.$$

It follows similarly that

$$\left(1 + i\varepsilon\Omega - \tfrac{1}{2}\varepsilon^2\Omega^2\right) Q_2 \left(1 - i\varepsilon\Omega - \tfrac{1}{2}\varepsilon^2\Omega^2\right) = Q_2 + \varepsilon \frac{1}{m} P_2,$$

$$\left(1 + i\varepsilon\Omega - \tfrac{1}{2}\varepsilon^2\Omega^2\right) Q_3 \left(1 - i\varepsilon\Omega - \tfrac{1}{2}\varepsilon^2\Omega^2\right) = Q_3 + \varepsilon \frac{1}{m} P_3.$$

All these equations are satisfied if

$$\Omega = \frac{1}{\hbar} H$$

with

$$H = \frac{1}{2m}\left(P_1^2 + P_2^2 + P_3^2\right),$$

$$= \tfrac{1}{2} m\left(V_1^2 + V_2^2 + V_3^2\right),$$

which is the matrix that represents the energy. We can see that P_1, P_2, P_3 are not changed because these matrices H and Ω commute with P_1, P_2, P_3. To see this formula for H gives the right change in time for Q_1, for example, we use the commutation relations of Q_1 with P_1, P_2, P_3 to calculate

$$Q_1\left(P_1^2 + P_2^2 + P_3^2\right) - \left(P_1^2 + P_2^2 + P_3^2\right)Q_1 = Q_1 P_1 P_1 - P_1 P_1 Q_1$$

$$= P_1 Q_1 P_1 + i\hbar P_1 - P_1 Q_1 P_1 - P_1(-i\hbar) = 2i\hbar P_1.$$

Thus we find that

$$Q_1 H - H Q_1 = i\frac{\hbar}{m} P_1$$

and

$$Q_1\Omega - \Omega Q_1 = i\frac{1}{m}P_1,$$

so, keeping only terms with ε and ε^2 as usual, we get

$$\left(1 + i\varepsilon\Omega - \tfrac{1}{2}\varepsilon^2\Omega^2\right)Q_1\left(1 - i\varepsilon\Omega - \tfrac{1}{2}\varepsilon^2\Omega^2\right)$$

$$= Q_1 - i\varepsilon(Q_1\Omega - \Omega Q_1) - \tfrac{1}{2}\varepsilon^2[(Q_1\Omega - \Omega Q_1)\Omega - \Omega(Q_1\Omega - \Omega Q_1)]$$

$$= Q_1 - i\varepsilon i\frac{1}{m}P_1 - \tfrac{1}{2}\varepsilon^2\left[i\frac{1}{m}P_1\Omega - \Omega i\frac{1}{m}P_1\right] = Q_1 + \varepsilon\frac{1}{m}P_1.$$

We can see similarly that these matrices H and Ω give the right changes in time for Q_2 and Q_3.

This matrix Ω commutes, as it should, with the matrices

$$K_1 = \frac{1}{\hbar}P_1, \qquad K_2 = \frac{1}{\hbar}P_2, \qquad K_3 = \frac{1}{\hbar}P_3$$

and

$$J_1 = \frac{1}{\hbar}L_1, \qquad J_2 = \frac{1}{\hbar}L_2, \qquad J_3 = \frac{1}{\hbar}L_3$$

for translations and rotations. For J_1, J_2, J_3 this was demonstrated in Problem 20-3.

Now suppose the system consists of two particles, for example an electron and an atomic nucleus. Let the position and momentum of the electron be represented by matrices Q_{e1}, Q_{e2}, Q_{e3} and P_{e1}, P_{e2}, P_{e3} and the position and momentum of the nucleus by matrices Q_{n1}, Q_{n2}, Q_{n3} and P_{n1}, P_{n2}, P_{n3}. We assume, as in Chapter 20, that the matrices for each particle satisfy the usual commutation relations and all the matrices for the electron commute with all the matrices for the nucleus. Then the matrices

$$K_1 = \frac{1}{\hbar}(P_{e1} + P_{n1}),$$

$$K_2 = \frac{1}{\hbar}(P_{e2} + P_{n2}),$$

$$K_3 = \frac{1}{\hbar}(P_{e3} + P_{n3})$$

give the right changes of the position and momentum matrices of both particles for space translations. Let L_{e1}, L_{e2}, L_{e3} be the matrices representing orbital angular momentum made from the position and momentum matrices for the electron and L_{n1}, L_{n2}, L_{n3} those made from the position and momentum matrices for the nucleus. Then the matrices

$$J_1 = \frac{1}{\hbar}(L_{e1} + L_{n1}),$$

$$J_2 = \frac{1}{\hbar}(L_{e2} + L_{n2}),$$

$$J_3 = \frac{1}{\hbar}(L_{e3} + L_{n3})$$

give the right changes of the position and momentum matrices of both particles for rotations. If there were no force between the particles and both moved with constant velocity, the right changes in time of the position and momentum matrices of both particles would be produced by the matrix

$$\Omega = \frac{1}{\hbar}H$$

with

$$H = \frac{1}{2m_e}(P_{e1}{}^2 + P_{e2}{}^2 + P_{e3}{}^2) + \frac{1}{2m_n}(P_{n1}{}^2 + P_{n2}{}^2 + P_{n3})^2$$

where m_e and m_n are the masses of the electron and nucleus. Without the factors of $1/\hbar$, these are the matrices that represent the total momentum, orbital angular momentum, and energy of motion of the two particles. They satisfy the right commutation relations with each other. The matrices K_1, K_2, K_3 commute with each other, reflecting the fact that translations in different directions commute. The matrices J_1, J_2, J_3 satisfy the commutation relations that correspond to multiplication of rotations around different axes. The matrices K_1, K_2, K_3 and J_1, J_2, J_3 satisfy the commutation relations that correspond to multiplication of translations and rotations. The matrix Ω commutes with K_1, K_2, K_3 and J_1, J_2, J_3, reflecting the fact that changes of the time coordinate commute with changes of the space coordinates. Everything is the same as for a single particle; it is just done twice.

A force between the two particles is included by letting

$$H = \frac{1}{2m_e}(P_{e1}{}^2 + P_{e2}{}^2 + P_{e3}{}^2) + \frac{1}{2m_n}(P_{n1}{}^2 + P_{n2}{}^2 + P_{n3}{}^2) + V,$$

where V is the matrix that represents the potential energy of the interaction between the two particles. Then H represents the total energy, which is the sum of the potential energy and energy of motion. For example, for the atom with one electron that we considered in Chapter 22,

$$V = -Ze^2R^{-1},$$

where R is the matrix that represents the distance between the electron and nucleus. Generally V does not commute with the matrices P_{e1}, P_{e2}, P_{e3} and P_{n1}, P_{n2}, P_{n3} that represent the momenta of the particles. Then H and Ω do not commute with P_{e1}, P_{e2}, P_{e3} and P_{n1}, P_{n2}, P_{n3}. This means the momenta are not constant in time; the velocities change; the particles are accelerated. However, K_1, K_2, K_3 and J_1, J_2, J_3 still must commute with Ω because changes of the space coordinates commute with changes of the time coordinate; the total momentum and angular momentum are constant in time and the total energy is not changed by space translations and rotations. This requires that V commutes with K_1, K_2, K_3 and J_1, J_2, J_3. Implications of these requirements will be discussed in the next chapter.

PROBLEM

26-1. Consider a single particle with a spin and magnetic moment described by matrices $\Sigma_1, \Sigma_2, \Sigma_3$ in addition to position and momentum represented by matrices Q_1, Q_2, Q_3 and P_1, P_2, P_3. Then K_1, K_2, K_3 are the same as in Problem 25-2 and J_1, J_2, J_3 are the same as in Problem 24-6. Let

$$H = \frac{1}{2m}\left(P_1^{\,2} + P_2^{\,2} + P_3^{\,2}\right).$$

Show that the matrix $\Omega = (1/\hbar)H$ commutes with K_1, K_2, K_3 and J_1, J_2, J_3. It also commutes with $\Sigma_1, \Sigma_2, \Sigma_3$, so the spin and magnetic moment are constant in time. Show that the orbital angular momentum represented by L_1, L_2, L_3 also is constant in time.

27 CHANGES IN VELOCITY

To complete the scheme we consider changes in velocity. Then we can see what all this implies for an interaction between two particles.

Suppose an object is moving in a certain direction with constant velocity v. In time t it moves a distance vt in that direction. If x_0 is its position coordinate in that direction at time $t = 0$, then at other times t its position coordinate is

$$x = x_0 + vt.$$

If its velocity were $v - \varepsilon$, its position coordinate at time t would be

$$\begin{aligned} x_0 + (v - \varepsilon)t &= x_0 + vt - \varepsilon t \\ &= x - \varepsilon t. \end{aligned}$$

The velocity $v - \varepsilon$ is the velocity of the object measured by observers who are moving in the same direction with velocity ε. The corresponding position coordinate is the one they measure, if they use the same time coordinate as we do and a reference point for the position coordinate that is the same as ours at $t = 0$. The change of coordinate relates descriptions of the same object by observers moving with velocity ε relative to each other.

We have to accommodate such changes. That is Newton's first law. The object may be at rest or moving with one constant velocity or another. It depends on the velocity of the observer, and the descriptions by observers moving at different velocities are equivalent. This idea comes from Galileo Galilei, so the change of coordinate is called a Galilei transformation.

We consider the change in the description of a system in quantum mechanics made by this change of coordinate. Let P be the matrix that represents the momentum and let m be the mass of the object. The velocity is represented by the matrix

$$V = \frac{1}{m} P.$$

It is changed to $V - \varepsilon$, so P is changed to

$$P - \varepsilon m.$$

Let Q be the matrix that represents the position coordinate at time $t = 0$. It is not changed; there is no change in the coordinate at $t = 0$. The matrix that represents the position coordinate at time t is

$$Q + tV = Q + \frac{t}{m} P.$$

It is changed, by the change of P, to

$$Q + \frac{t}{m}(P - \varepsilon m) = Q + t(V - \varepsilon) = (Q + tV) - \varepsilon t.$$

For velocities ε that are small in some suitable units, this change is generally made by multiplying with a matrix

$$1 - i\varepsilon G - \tfrac{1}{2}\varepsilon^2 G^2$$

and its inverse

$$1 + i\varepsilon G - \tfrac{1}{2}\varepsilon^2 G^2,$$

changing P, for example, to

$$\begin{aligned}&(1 + i\varepsilon G - \tfrac{1}{2}\varepsilon^2 G^2) P (1 - i\varepsilon G - \tfrac{1}{2}\varepsilon^2 G^2)\\ &\quad = P - i\varepsilon(PG - GP) - \tfrac{1}{2}\varepsilon^2[(PG - GP)G - G(PG - GP)].\end{aligned}$$

Again, the right side is obtained by multiplying out the product on the left. As usual, we keep only terms with ε and ε^2. For the result to be $P - \varepsilon m$ for any choice of ε, we must have

$$PG - GP = -im.$$

Then the term with ε^2 is zero and we do get

$$(1 + i\varepsilon G - \tfrac{1}{2}\varepsilon^2 G^2) P (1 - i\varepsilon G - \tfrac{1}{2}\varepsilon^2 G^2) = P - \varepsilon m.$$

To have Q not changed we must have

$$QG - GQ = 0.$$

Then we get

$$\left(1 + i\varepsilon G - \tfrac{1}{2}\varepsilon^2 G^2\right)Q\left(1 - i\varepsilon G - \tfrac{1}{2}\varepsilon^2 G^2\right) = Q.$$

These equations are satisfied by letting

$$G = \frac{m}{\hbar}Q,$$

provided Q and P satisfy the commutation relation

$$QP - PQ = i\hbar.$$

This gives further meaning to Born's "strange equation." The matrix Q also plays two roles. In addition to representing the position coordinate, it provides the matrix G for Galilei transformations.

Now we extend this to three-dimensional position and momentum. Consider a Galilei transformation in the 1 direction, for example, which means a change of the velocity in the 1 direction. The matrix V_1 representing the velocity in the 1 direction is changed to $V_1 - \varepsilon$, so the matrix P_1 representing the momentum in the 1 direction is changed to $P_1 - \varepsilon m$. The velocities in the 2 and 3 directions are not changed, so the matrices P_2 and P_3 that represent momentum in the 2 and 3 directions are not changed. Let Q_1, Q_2, Q_3 be the matrices that represent position coordinates at time $t = 0$. They are not changed. For small velocities ε, this change is generally made with a matrix G_1 that gives

$$\left(1 + i\varepsilon G_1 - \tfrac{1}{2}\varepsilon^2 G_1^{\,2}\right)Q_1\left(1 - i\varepsilon G_1 - \tfrac{1}{2}\varepsilon^2 G_1^{\,2}\right) = Q_1,$$

$$\left(1 + i\varepsilon G_1 - \tfrac{1}{2}\varepsilon^2 G_1^{\,2}\right)Q_2\left(1 - i\varepsilon G_1 - \tfrac{1}{2}\varepsilon^2 G_1^{\,2}\right) = Q_2,$$

$$\left(1 + i\varepsilon G_1 - \tfrac{1}{2}\varepsilon^2 G_1^{\,2}\right)Q_3\left(1 - i\varepsilon G_1 - \tfrac{1}{2}\varepsilon^2 G_1^{\,2}\right) = Q_3,$$

$$\left(1 + i\varepsilon G_1 - \tfrac{1}{2}\varepsilon^2 G_1^{\,2}\right)P_1\left(1 - i\varepsilon G_1 - \tfrac{1}{2}\varepsilon^2 G_1^{\,2}\right) = P_1 - \varepsilon m,$$

$$\left(1 + i\varepsilon G_1 - \tfrac{1}{2}\varepsilon^2 G_1^{\,2}\right)P_2\left(1 - i\varepsilon G_1 - \tfrac{1}{2}\varepsilon^2 G_1^{\,2}\right) = P_2,$$

$$\left(1 + i\varepsilon G_1 - \tfrac{1}{2}\varepsilon^2 G_1^{\,2}\right)P_3\left(1 - i\varepsilon G_1 - \tfrac{1}{2}\varepsilon^2 G_1^{\,2}\right) = P_3.$$

From our previous calculations, we know this means

$$Q_1G_1 - G_1Q_1 = 0,$$

$$Q_2G_1 - G_1Q_2 = 0,$$

$$Q_3G_1 - G_1Q_3 = 0,$$

$$P_1G_1 - G_1P_1 = -im,$$

$$P_2G_1 - G_1P_2 = 0,$$

$$P_3G_1 - G_1P_3 = 0.$$

Similarly, Galilei transformations in the 2 direction are represented with a matrix G_2 that commutes with Q_1, Q_2, Q_3 and P_1, P_3 and gives

$$P_2G_2 - G_2P_2 = -im,$$

and Galilei transformations in the 3 direction are represented with a matrix G_3 that commutes with Q_1, Q_2, Q_3 and P_1, P_2 and gives

$$P_3G_3 - G_3P_3 = -im.$$

Galilei transformations are included in the general scheme that follows from assumptions valid for any isolated system. They are combined with space translations, rotations, and changes in time by considering all of them as changes of space and time coordinates, so the product of any two of them is the change of coordinates made by doing first one and then the other. This gives commutation relations for G_1, G_2, G_3 with K_1, K_2, K_3; J_1, J_2, J_3, and Ω.

We are particularly interested in the commutation relations of G_1, G_2, G_3 with Ω. To find that of G_1 with Ω, for example, we consider the matrix product

$$\left(1 - i\varepsilon G_1 - \tfrac{1}{2}\varepsilon^2 G_1{}^2\right)\left(1 - i\varepsilon\Omega - \tfrac{1}{2}\varepsilon^2\Omega^2\right)$$

$$\times\left(1 + i\varepsilon G_1 - \tfrac{1}{2}\varepsilon^2 G_1{}^2\right)\left(1 + i\varepsilon\Omega - \tfrac{1}{2}\varepsilon^2\Omega^2\right) = 1 + \varepsilon^2(\Omega G_1 - G_1\Omega).$$

Again, the right side is obtained by working out the product on the left side, as we did in Chapter 24 for rotations. Now the four matrices in the product represent a Galilei transformation in the 1 direction, a change in time, and their inverses. The matrix

$$1 - i\varepsilon G_1 - \tfrac{1}{2}\varepsilon^2 G_1{}^2$$

corresponds to the change of the space and time coordinates x_1 and t to $x_1 - \varepsilon t$ and t. The matrix

$$1 - i\varepsilon\Omega - \tfrac{1}{2}\varepsilon^2\Omega^2$$

corresponds to the change of the time coordinate from t to $t - \varepsilon$ because a matrix B representing a quantity at the time you call noon is changed to

$$\left(1 + i\varepsilon\Omega - \tfrac{1}{2}\varepsilon^2\Omega^2\right)B\left(1 - i\varepsilon\Omega - \tfrac{1}{2}\varepsilon^2\Omega^2\right)$$

to represent the quantity at the time I call noon if my clock lags behind yours by ε, so when your time is t mine is $t - \varepsilon$. Since x_2 and x_3 are not changed, we find the product of the four changes of coordinates by computing the changes of x_1 and t in succession, just as we did with x_1 and x_2 to find the commutation relation for J_3 and K_2 in Chapter 25.

First there is the Galilei transformation in the 1 direction, which changes x_1 and t to

$$x_1 - \varepsilon t \quad \text{and} \quad t.$$

Then the time t is changed to $t - \varepsilon$ in these, which gives

$$x_1 - \varepsilon(t - \varepsilon) \quad \text{and} \quad t - \varepsilon.$$

Next there is the inverse Galilei transformation, which changes x_1 and t in these to $x_1 + \varepsilon t$ and t and gives

$$x_1 + \varepsilon t - \varepsilon(t - \varepsilon) \quad \text{and} \quad t - \varepsilon.$$

Finally, there is the inverse of the change in time, from t to $t + \varepsilon$, which gives

$$x_1 + \varepsilon(t + \varepsilon) - \varepsilon(t + \varepsilon - \varepsilon) = x_1 + \varepsilon t + \varepsilon^2 - \varepsilon t = x_1 + \varepsilon^2$$

and

$$t + \varepsilon - \varepsilon = t.$$

This is the result of the four changes of coordinates. Altogether x_1 is changed to $x_1 + \varepsilon^2$ and t is not changed. Since x_2 and x_3 also are not changed, the product of the four changes of coordinates is just a translation a distance ε^2 in the 1 direction. It is represented by the matrix

$$1 - i\varepsilon^2 K_1.$$

Then

$$1 + \varepsilon^2(\Omega G_1 - G_1\Omega) = 1 - i\varepsilon^2 K_1$$

because the product of the four matrices is the matrix that represents the product of the four changes of coordinates. This is part of the general scheme that follows from assumptions valid for any isolated system. The last equation is not an approximation. It has to hold when higher powers of ε are included because the full equation holds for any choice of the small number ε. Therefore

$$\Omega G_1 - G_1\Omega = -iK_1.$$

It follows similarly that

$$\Omega G_2 - G_2\Omega = -iK_2$$

and

$$\Omega G_3 - G_3\Omega = -iK_3.$$

Consider a single particle with position coordinates and momentum represented by matrices Q_1, Q_2, Q_3 and P_1, P_2, P_3. Let Ω and K_1, K_2, K_3 be the same as before. The commutation relations of G_1, G_2, G_3 with Q_1, Q_2, Q_3; P_1, P_2, P_3; and Ω are satisfied by letting

$$G_1 = \frac{m}{\hbar}Q_1, \qquad G_2 = \frac{m}{\hbar}Q_2, \qquad G_3 = \frac{m}{\hbar}Q_3.$$

In particular, our calculation of

$$Q_1\Omega - \Omega Q_1 = i\frac{1}{m}P_1$$

in the last chapter shows that the commutation relations of G_1, G_2, G_3 with Ω are satisfied. Writing them in the form of the last equation shows these commutation relations correspond to the way the matrices representing position coordinates are changed in time. They also correspond to the way the matrix representing energy is changed by Galilei transformations. That is to be developed in Problem 27-5.

Now consider the system of two particles with K_1, K_2, K_3; J_1, J_2, J_3; and Ω the same as in the last chapter.

Let

$$G_1 = \frac{m_e}{\hbar} Q_{e1} + \frac{m_n}{\hbar} Q_{n1},$$

$$G_2 = \frac{m_e}{\hbar} Q_{e2} + \frac{m_n}{\hbar} Q_{n2},$$

$$G_3 = \frac{m_e}{\hbar} Q_{e3} + \frac{m_n}{\hbar} Q_{n3}.$$

These matrices G_1, G_2, G_3 are chosen so that with the position and momentum matrices for both particles, they satisfy the commutation relations that follow from the way position coordinates and momenta are changed by Galilei transformations. If there is no force between the particles, and thus no matrix V, the commutation relations of G_1, G_2, G_3 with Ω also are satisfied. Again, all this is the same as for a single particle; it is just done twice.

When there is a force between the particles, the commutation relations of G_1, G_2, G_3 with Ω require that V commutes with G_1, G_2, G_3. Then V commutes with K_1, K_2, K_3; J_1, J_2, J_3; and G_1, G_2, G_3. Now we shall see what this implies.

The matrices

$$R_1 = \frac{\hbar}{m_e + m_n} G_1 = \frac{m_e}{m_e + m_n} Q_{e1} + \frac{m_n}{m_e + m_n} Q_{n1},$$

$$R_2 = \frac{\hbar}{m_e + m_n} G_2 = \frac{m_e}{m_e + m_n} Q_{e2} + \frac{m_n}{m_e + m_n} Q_{n2},$$

$$R_3 = \frac{\hbar}{m_e + m_n} G_3 = \frac{m_e}{m_e + m_n} Q_{e3} + \frac{m_n}{m_e + m_n} Q_{n3}$$

represent the position coordinates of the center of mass of the two particles. The matrices $\hbar K_1, \hbar K_2, \hbar K_3$ represent the total momentum of the two particles. Matrices Q_1, Q_2, Q_3 and P_1, P_2, P_3 representing the relative position and momentum were introduced in Chapter 20. Now we use this set of 12 matrices instead of the set of 12 matrices that represent the position coordinates and momenta of the individual particles. Each set can be written in terms of the other. We have equations that give the former in terms of the latter. It is easy to solve these equations and get formulas for the latter in terms of the former. That is to be done in Problem 27-1.

All the matrices R_1, R_2, R_3 and $\hbar K_1, \hbar K_2, \hbar K_3$ commute with all the matrices Q_1, Q_2, Q_3 and P_1, P_2, P_3. That is to be shown in Problem 27-2. For $\hbar G_1$ and $\hbar K_1$ we have the commutation relation

$$\begin{aligned} &\hbar G_1 \hbar K_1 - \hbar K_1 \hbar G_1 \\ &\quad = (m_e Q_{e1} + m_n Q_{n1})(P_{e1} + P_{n1}) - (P_{e1} + P_{n1})(m_e Q_{e1} + m_n Q_{n1}) \\ &\quad = m_e(Q_{e1}P_{e1} - P_{e1}Q_{e1}) + m_n(Q_{n1}P_{n1} - P_{n1}Q_{n1}) = i\hbar(m_e + m_n), \end{aligned}$$

so for R_1 and K_1 we have

$$R_1 K_1 - K_1 R_1 = i$$

and for $\hbar K_1$ and G_1 we have

$$\hbar K_1 G_1 - G_1 \hbar K_1 = -i(m_e + m_n).$$

From calculations like several we have done before, it follows that

$$\left(1 + i\varepsilon K_1 - \tfrac{1}{2}\varepsilon^2 K_1{}^2\right) R_1 \left(1 - i\varepsilon K_1 - \tfrac{1}{2}\varepsilon^2 K_1{}^2\right) = R_1 + \varepsilon,$$

which is correct for the change of the position coordinate by the space translation, and

$$\left(1 + i\varepsilon G_1 - \tfrac{1}{2}\varepsilon^2 G_1{}^2\right) \hbar K_1 \left(1 - i\varepsilon G_1 - \tfrac{1}{2}\varepsilon^2 G_1{}^2\right) = \hbar K_1 - \varepsilon(m_e + m_n),$$

which is correct for the change of the total momentum by the Galilei transformation.

Suppose V is made from the matrices that represent position coordinates and momenta. Then V can be made from R_1, R_2, R_3; $\hbar K_1, \hbar K_2, \hbar K_3$; and Q_1, Q_2, Q_3; P_1, P_2, P_3. The requirement that V commutes with K_1, K_2, K_3 and G_1, G_2, G_3 implies that V is made only from Q_1, Q_2, Q_3 and P_1, P_2, P_3. To see this, consider what happens when V is changed to

$$\left(1 + i\varepsilon K_1 - \tfrac{1}{2}\varepsilon^2 K_1{}^2\right) V \left(1 - i\varepsilon K_1 - \tfrac{1}{2}\varepsilon^2 K_1{}^2\right).$$

Since K_1 commutes with all the matrices V can be made from except R_1, the result is just that R_1 in V is changed to $R_1 + \varepsilon$. That happens if R_1 occurs in V multiplied by a number, in a sum or product of matrices, in a square or inverse of a matrix, or in any combination of these. That was demonstrated in Problem 23-1. It also happens if R_1 occurs in a square root. What V is made from is all that is important here. How it is made is not important. On the other hand, since V must commute with K_1, the result must also be that

$$\left(1 + i\varepsilon K_1 - \tfrac{1}{2}\varepsilon^2 K_1^{\,2}\right)V\left(1 - i\varepsilon K_1 - \tfrac{1}{2}\varepsilon^2 K_1^{\,2}\right) = V.$$

Therefore V is the same when R_1 in it is changed to $R_1 + \varepsilon$ for any small ε. We conclude there is no R_1 in V. Similarly, since V must commute with G_1, we conclude there is no $\hbar K_1$ in V. Since V must also commute with K_2, K_3 and G_2, G_3, it follows similarly that V does not depend on R_2, R_3 or $\hbar K_2, \hbar K_3$. Then V has to be made only from the matrices Q_1, Q_2, Q_3 and P_1, P_2, P_3 that represent the relative position coordinates and momenta.

The remaining requirement is that V commutes with J_1, J_2, J_3. For these we have

$$\hbar J_1 = R_2 \hbar K_3 - R_3 \hbar K_2 + L_1,$$

$$\hbar J_2 = R_3 \hbar K_1 - R_1 \hbar K_3 + L_2,$$

$$\hbar J_3 = R_1 \hbar K_2 - R_2 \hbar K_1 + L_3.$$

This is to be shown in Problem 27-3. The first parts are the matrices representing the orbital angular momentum made from the total momentum and the position coordinates of the center of mass. As before, L_1, L_2, L_3 are the matrices representing the orbital angular momentum made from the relative position and momentum. Since V commutes with R_1, R_2, R_3 and K_1, K_2, K_3 as well as with J_1, J_2, J_3, it follows that V commutes with L_1, L_2, L_3.

The matrix H that represents the energy can be written as

$$H = \frac{\hbar^2}{2(m_e + m_n)}\left(K_1^2 + K_2^2 + K_3^2\right) + \frac{1}{2m}\left(P_1^2 + P_2^2 + P_3^2\right) + V,$$

where, as before,

$$m = \frac{m_e m_n}{m_e + m_n}.$$

This is to be shown in Problem 27-4. In terms of the matrices

$$W_j = \frac{m_e}{m_e + m_n} V_{ej} + \frac{m_n}{m_e + m_n} V_{nj}$$

$$= \frac{1}{m_e + m_n}\left(P_{ej} + P_{nj}\right) = \frac{\hbar}{m_e + m_n} K_j$$

for $j = 1, 2, 3$, which represent the velocity of the center of mass, the first part of H is

$$\tfrac{1}{2}(m_e + m_n)(W_1^2 + W_2^2 + W_3^2).$$

It represents the energy of motion of the total mass $m_e + m_n$ moving with the velocity of the center of mass. That is the energy of motion of the atom or system of two particles as a whole. The rest of H represents the energy of the two particles in the system. This is what we used in Chapter 22 to represent the energy of an electron and nucleus in an atom. In general, for any V that satisfies the requirements, this part of H depends only on the relative position and momentum of the two particles. It commutes with L_1, L_2, L_3 because both $P_1^2 + P_2^2 + P_3^2$ and V commute with L_1, L_2, L_3. We used this in Chapter 22 to find the possible values for the energy in an atom.

The requirements that V commutes with G_1, G_2, G_3; K_1, K_2, K_3; and J_1, J_2, J_3 follow from assumptions that are valid for an isolated system, so they are applied to the entire system of two particles.

PROBLEMS

27-1. Starting with the equations for R_1, R_2, R_3 in this chapter, those for K_1, K_2, K_3 for two particles from the last chapter, and the equations for the relative position and momentum matrices Q_1, Q_2, Q_3 and P_1, P_2, P_3 given in Chapter 20, find the formulas for Q_{e1}, Q_{e2}, Q_{e3}; P_{e1}, P_{e2}, P_{e3} and Q_{n1}, Q_{n2}, Q_{n3}; P_{n1}, P_{n2}, P_{n3} in terms of R_1, R_2, R_3; $\hbar K_1, \hbar K_2, \hbar K_3$ and Q_1, Q_2, Q_3; P_1, P_2, P_3. You can check that your answers are correct by substituting them into the equations you started with and seeing that those equations are satisfied.

27-2. Show that for two particles all the matrices R_1, R_2, R_3 and $\hbar K_1, \hbar K_2, \hbar K_3$ commute with all the matrices Q_1, Q_2, Q_3 and P_1, P_2, P_3 that represent the relative position and momentum. Use the same equations you used for the last problem. Assume that the position and momentum matrices for each particle satisfy the usual commutation relations and that all the matrices for one particle commute with all the matrices for the other.

27-3. Show that for two particles

$$\hbar J_1 = R_2 \hbar K_3 - R_3 \hbar K_2 + Q_2 P_3 - Q_3 P_2,$$

$$\hbar J_2 = R_3 \hbar K_1 - R_1 \hbar K_3 + Q_3 P_1 - Q_1 P_3,$$

$$\hbar J_3 = R_1 \hbar K_2 - R_2 \hbar K_1 + Q_1 P_2 - Q_2 P_1,$$

where all the matrices on the right sides are the same as in the last two problems. You can do this two different ways. One way is to substitute the results of the first problem into the formulas for $\hbar J_1, \hbar J_2, \hbar J_3$ in terms of the position and momentum matrices of the individual particles. The other way is to work backward, substitute the formulas you used in the first problem for R_1, R_2, R_3; $\hbar K_1, \hbar K_2, \hbar K_3$ and Q_1, Q_2, Q_3; P_1, P_2, P_3 into the preceding formulas for $\hbar J_1, \hbar J_2, \hbar J_3$, and see that you get the correct formulas in terms of the position and momentum matrices of the individual particles.

27-4. Show that for two particles

$$\frac{1}{2m_e}\left(P_{e1}{}^2 + P_{e2}{}^2 + P_{e3}{}^2\right) + \frac{1}{2m_n}\left(P_{n1}{}^2 + P_{n2}{}^2 + P_{n3}{}^2\right)$$

$$= \frac{\hbar^2}{2(m_e + m_n)}\left(K_1{}^2 + K_2{}^2 + K_3{}^2\right) + \frac{1}{2m}\left(P_1{}^2 + P_2{}^2 + P_3{}^2\right),$$

where all the matrices are the same as in the last three problems and

$$m = \frac{m_e m_n}{m_e + m_n}$$

as before. This can be done two different ways the same as the last problem.

27-5. Let

$$U = 1 - i\varepsilon G_1 - \tfrac{1}{2}\varepsilon^2 G_1{}^2$$

and calculate $U^{-1}HU$, where

$$H = \frac{1}{2m}\left(P_1{}^2 + P_2{}^2 + P_3{}^2\right)$$

for a single particle. Do this two different ways. First use this formula for H, the formulas for $U^{-1}P_1U$, $U^{-1}P_2U$, $U^{-1}P_3U$, and the fact that for each matrix P

$$U^{-1}P^2U = U^{-1}PUU^{-1}PU = \left(U^{-1}PU\right)^2.$$

Then use

$$U^{-1}HU = H - i\varepsilon(HG_1 - G_1H)$$
$$- \tfrac{1}{2}\varepsilon^2[(HG_1 - G_1H)G_1 - G_1(HG_1 - G_1H)],$$

the commutation relation

$$\Omega G_1 - G_1\Omega = -iK_1,$$

which can be written here as

$$HG_1 - G_1H = -iP_1,$$

and the commutation relation

$$P_1G_1 - G_1P_1 = -im,$$

which corresponds to the way P_1 is changed by Galilei transformations. If you get the same answer, it shows the commutation relation of Ω and G_1 does correspond to the way energy is changed by a Galilei transformation.

27-6. Matrices V are proposed to describe various interactions between two particles. Which of the following matrices V satisfy the requirements that V commutes with G_1, G_2, G_3; K_1, K_2, K_3; and J_1, J_2, J_3? For each that fails to satisfy the requirements, explain where it fails.

(a) $V = R_1{}^2 + R_2{}^2 + R_3{}^2$ (e) $V = L_3$
(b) $V = K_1{}^2 + K_2{}^2 + K_3{}^2$ (f) $V = Q_3$
(c) $V = Q_1{}^2 + Q_2{}^2 + Q_3{}^2$ (g) $V = Q_1L_1 + Q_2L_2 + Q_3L_3$
(d) $V = P_1{}^2 + P_2{}^2 + P_3{}^2$ (h) $V = Q_1R_1 + Q_2R_2 + Q_3R_3$

All the matrices in these are the same as in the first four problems. The results of Problems 24-3 and 24-4 can be used here.

27-7. Consider a system of two particles with spins and magnetic moments described by matrices $\Sigma_1, \Sigma_2, \Sigma_3$ and Ξ_1, Ξ_2, Ξ_3 as in Chapter 14. Let the position coordinates and momenta be represented by matrices Q_{e1}, Q_{e2}, Q_{e3}; P_{e1}, P_{e2}, P_{e3} and Q_{n1}, Q_{n2}, Q_{n3}; P_{n1}, P_{n2}, P_{n3} as before. All the matrices for one particle commute with all the matrices for the other. For each particle, the matrices are like those in Problems 24-6, 25-2, and 26-1. Let H, Ω, K_1, K_2, K_3 and G_1, G_2, G_3

for the two particles be the same as before, and let

$$J_1 = \frac{1}{\hbar} L_{e1} + \frac{1}{\hbar} L_{n1} + \tfrac{1}{2}\Sigma_1 + \tfrac{1}{2}\Xi_1,$$

$$J_2 = \frac{1}{\hbar} L_{e2} + \frac{1}{\hbar} L_{n2} + \tfrac{1}{2}\Sigma_2 + \tfrac{1}{2}\Xi_2,$$

$$J_3 = \frac{1}{\hbar} L_{e3} + \frac{1}{\hbar} L_{n3} + \tfrac{1}{2}\Sigma_3 + \tfrac{1}{2}\Xi_3,$$

where L_{e1}, L_{e2}, L_{e3} and L_{n1}, L_{n2}, L_{n3} are the matrices that represent the orbital angular momenta of the two particles, the same as in the last chapter. Again, matrices V are proposed to describe various interactions between the particles. Which of the following matrices V satisfy the requirements that V commutes with G_1, G_2, G_3; K_1, K_2, K_3; and J_1, J_2, J_3? For each that fails to satisfy the requirements, explain where it fails.

(a) $V = \Sigma_1\Xi_1 + \Sigma_2\Xi_2 + \Sigma_3\Xi_3$
(b) $V = R_1\Sigma_1 + R_2\Sigma_2 + R_3\Sigma_3$
(c) $V = K_1\Xi_1 + K_2\Xi_2 + K_3\Xi_3$
(d) $V = Q_1\Sigma_1 + Q_2\Sigma_2 + Q_3\Sigma_3$
(e) $V = \Sigma_3$
(f) $V = \Sigma_3\Xi_3$

The matrices Q_1, Q_2, Q_3 and R_1, R_2, R_3 are the same as in the first four problems. Again, the result of Problem 24-4 can be used here.

28 INVARIANCE AND WHAT IT IMPLIES

The whole scheme of matrices representing changes of space and time coordinates follows from rather simple assumptions. They apply quite generally to any isolated system. Now we consider what these assumptions are and what they imply. We find they imply almost all the commutation relations. We do not have to assume the commutation relations of position matrices with momentum matrices. We do not have to begin with them. We begin instead with the matrices that represent changes of coordinates. Their commutation relations follow from corresponding multiplications of changes of coordinates. From them we can deduce the commutation relations of position matrices with momentum matrices. This confirms that we have found the origin of Born's "strange equation."

The assumptions are about the changes in the description of a system in quantum mechanics made by changes of the space and time coordinates. The idea is that these changes do not make much difference. What is important is not changed. It is invariant. What is changed is not more important than arbitrary choices of coordinates.

It is assumed that when we make a change of coordinates we change a description of the system into an equivalent description. It is assumed that each of these descriptions can be obtained from the other; we can change the second description back into the first by making the inverse change of coordinates. Both descriptions say the same thing. They just say it in terms of different coordinates.

Suppose we make two changes, one after the other. It is assumed the second change is made the same as if it were first. The way we make the second change does not depend on the change we make before it.

This is assumed for rotations, space translations, changes in time, Galilei transformations, and all their products. For example, it is assumed that the way the description changes in time is not changed by changes in time or by rotations, space translations, or Galilei transformations. The dynamics is invariant in all these ways. That means there are no forces that depend on time, on the orientation or location of the system in space, or on its velocity.

These assumptions are valid for an isolated system that is not influenced by the times of events outside it or the orientation, location, or velocity of the system relative to anything outside it. For example, they apply to an atom that is isolated, but not to part of an atom; the force on an electron in an atom depends on its location relative to the nucleus.

From these and minor technical assumptions, it can be shown that the changes in the description of a system in quantum mechanics can be made with matrices J_1, J_2, J_3; K_1, K_2, K_3; Ω and G_1, G_2, G_3 as in the last four chapters [1–8]. It can be shown that generally a product of matrices that represent changes of coordinates is the matrix that represents the product of the changes of coordinates. This is true for all the products we have used to find commutation relations. There will be one exception, which will be discussed.

There are a few more commutation relations for us to consider. The matrices G_1, G_2, G_3 commute with each other. This is obtained the same way as that K_1, K_2, K_3 commute with each other, as discussed in Chapter 25. It reflects the fact that Galilei transformations in different directions commute. It is true for the particular matrices G_1, G_2, G_3 that we have used; that is to be shown in Problem 28-1.

We get

$$J_3G_2 - G_2J_3 = -iG_1$$

the same way we got

$$J_3K_2 - K_2J_3 = -iK_1$$

in Chapter 25. The only differences are that K_2 is replaced by G_2, so the space translation that changes x_2 to $x_2 + \varepsilon$ is replaced by the Galilei transformation that changes x_2 to $x_2 - \varepsilon t$, and then the product of the four changes of coordinates is the Galilei transformation that changes x_1 to $x_1 - \varepsilon^2 t$ instead of the space translation that changes x_1 to $x_1 + \varepsilon^2$, which means K_1 is replaced by G_1. It follows similarly that

$$J_1G_2 - G_2J_1 = iG_3, \qquad J_2G_1 - G_1J_2 = -iG_3,$$

$$J_2G_3 - G_3J_2 = iG_1,$$

$$J_3G_1 - G_1J_3 = iG_2, \qquad J_1G_3 - G_3J_1 = -iG_2,$$

$$J_1G_1 - G_1J_1 = 0, \qquad J_2G_2 - G_2J_2 = 0,$$

$$J_3G_3 - G_3J_3 = 0.$$

We see that these commutation relations correspond to multiplications of Galilei transformations and rotations. They are satisfied by the particular matrices G_1, G_2, G_3 and J_1, J_2, J_3 that we have used; that is to be shown in Problem 28-2.

We get

$$G_1K_2 - K_2G_1 = 0, \qquad G_1K_3 - K_3G_1 = 0,$$

$$G_2K_1 - K_1G_2 = 0, \qquad G_2K_3 - K_3G_2 = 0,$$

$$G_3K_1 - K_1G_3 = 0, \qquad G_3K_2 - K_2G_3 = 0$$

the same way we got that K_1, K_2, K_3 commute with each other. These commutation relations reflect the fact that Galilei transformations commute with space translations. They are satisfied by the particular matrices G_1, G_2, G_3 and K_1, K_2, K_3 that we have used; that is to be shown in Problem 28-3.

However, in general

$$G_1K_1 - K_1G_1 = i\lambda,$$

$$G_2K_2 - K_2G_2 = i\lambda,$$

$$G_3K_3 - K_3G_3 = i\lambda,$$

where λ is a real number. These commutation relations are obtained as part of the general scheme that follows from assumptions valid for any isolated system. They are satisfied by the particular matrices G_1, G_2, G_3 and K_1, K_2, K_3 that we have used; for a single particle λ is $m/\hbar$ and for two particles λ is $(m_e + m_n)/\hbar$; that is to be found in Problem 28-3.

Galilei transformations commute with space translations. Then why does G_1 not commute with K_1? That has to be explained.

Here is the one exception to the general rule that a product of matrices that represent changes of coordinates is the matrix that represents the product of the changes of coordinates. Consider the matrix product

$$\left(1 - i\varepsilon K_1 - \tfrac{1}{2}\varepsilon^2 K_1{}^2\right)\left(1 - i\varepsilon G_1 - \tfrac{1}{2}\varepsilon^2 G_1{}^2\right)$$

$$\times\left(1 + i\varepsilon K_1 - \tfrac{1}{2}\varepsilon^2 K_1{}^2\right)\left(1 + i\varepsilon G_1 - \tfrac{1}{2}\varepsilon^2 G_1{}^2\right) = 1 + \varepsilon^2(G_1K_1 - K_1G_1).$$

As usual, we get the right side by working out the product on the left. We keep only terms with ε and ε^2. Substituting the commutation relation into the last equation gives

$$1 + i\varepsilon^2\lambda$$

for the matrix product. On the other hand, the four matrices in the product represent a space translation, a Galilei transformation, and their inverses. Since these four changes of coordinates commute, we can multiply them in any order. Their product is the identity change of coordinates, which is no change at all. It is represented by the matrix 1. The matrix that represents the product of the four changes of coordinates is 1, but the product of the four matrices is not. They are different, but they have the same effect in changing a description of the system. Changing a matrix B to

$$(1 + i\varepsilon^2\lambda)^{-1}B(1 + i\varepsilon^2\lambda) = B$$

is the same as changing it to

$$1 \cdot B \cdot 1 = B.$$

Thus $1 + i\varepsilon^2\lambda$ represents the identity change of coordinates as well as 1 does. The product of the four matrices does represent the product of the four changes of coordinates. That is all that is really needed. Remember that in Chapter 23 the identity rotation was represented by -1 as well as by 1.

There is some flexibility in the whole scheme. For example, we could change K_1 to $K_1 + b$ for any real number b. Consider the matrix

$$\begin{aligned} 1 - i\varepsilon(K_1 + b) - \tfrac{1}{2}\varepsilon^2(K_1 + b)^2 \\ = \left(1 - i\varepsilon K_1 - \tfrac{1}{2}\varepsilon^2 K_1{}^2\right)\left(1 - i\varepsilon b - \tfrac{1}{2}\varepsilon^2 b^2\right). \end{aligned}$$

We can see the right side is the same as the left by working out the product, keeping only terms with ε and ε^2 as usual. This matrix represents the space translation in the 1 direction as well as

$$1 - i\varepsilon K_1 - \tfrac{1}{2}\varepsilon^2 K_1{}^2$$

does, because it changes each matrix B to

$$\begin{aligned} &\left(1 - i\varepsilon b - \tfrac{1}{2}\varepsilon^2 b^2\right)^{-1}\left(1 - i\varepsilon K_1 - \tfrac{1}{2}\varepsilon^2 K_1{}^2\right)^{-1} \\ &\quad \times B\left(1 - i\varepsilon K_1 - \tfrac{1}{2}\varepsilon^2 K_1{}^2\right)\left(1 - i\varepsilon b - \tfrac{1}{2}\varepsilon^2 b^2\right) \\ &= \left(1 - i\varepsilon K_1 - \tfrac{1}{2}\varepsilon^2 K_1{}^2\right)^{-1} B\left(1 - i\varepsilon K_1 - \tfrac{1}{2}\varepsilon^2 K_1{}^2\right). \end{aligned}$$

Changing K_1 to $K_1 + b$ would change the commutation relation

$$\Omega G_1 - G_1\Omega = -iK_1.$$

It can be shown that with this flexibility of adding numbers, the matrices J_1, J_2, J_3; K_1, K_2, K_3; and G_1, G_2, G_3 can always be chosen to get the commutation relations we have described [1, 3, 6, 7]. We have seen that in particular cases the choice that works is the choice we would naturally make; we have no reason to add a number b to any of the particular matrices J_1, J_2, J_3; K_1, K_2, K_3; or G_1, G_2, G_3 that we have used. With this choice there is just one exception to the general rule that a product of matrices that represent changes of coordinates is the matrix that represents the product of the changes of coordinates. It holds for all the products we used to get commutation relations except those involving a Galilei transformation and a space translation in the same direction. There is not enough flexibility to get λ to be zero; in fact, it is not zero for the particular matrices we have used.

Now we have all the commutation relations for the matrices J_1, J_2, J_3; K_1, K_2, K_3; G_1, G_2, G_3; and Ω. From these we can deduce commutation relations for the matrices that represent the position coordinates and momentum of a particle. We use the way matrices representing position coordinates and momenta are changed by space translations and Galilei transformations. We also make an assumption to identify matrices that describe the particle. We do all this here for the description of a particle in one dimension. The same can be done for three dimensions. It shows the momentum matrices commute with each other and yields the commutation relations of position matrices with momentum matrices; that is to be done in Problem 28-4. For a particle with no spin, the same kind of argument shows the position matrices commute with each other; that is to be done in Problem 28-5. As demonstrated by Problem 28-6, it cannot show the position matrices commute with each other for a particle with spin; that requires another assumption [9].

For one dimension, there are space translations and Galilei transformations for only one direction, and no rotations, so changes of coordinates are represented with just three matrices K, G, and Ω. They satisfy the commutation relations

$$GK - KG = i\lambda,$$

$$G\Omega - \Omega G = iK,$$

$$K\Omega - \Omega K = 0,$$

where λ is a real number. Let

$$D = \Omega - \frac{1}{2\lambda}K^2.$$

Since $(1/2\lambda)K^2$ satisfies the commutation relations

$$G\frac{1}{2\lambda}K^2 - \frac{1}{2\lambda}K^2G = \frac{1}{2\lambda}(GK - KG)K + \frac{1}{2\lambda}K(GK - KG)$$

$$= \frac{1}{2\lambda}i\lambda K + \frac{1}{2\lambda}Ki\lambda = iK$$

and

$$K\frac{1}{2\lambda}K^2 - \frac{1}{2\lambda}K^2K = 0,$$

which are the same as the commutation relations of Ω with G and K, it follows that D must commute with G and K. Also $(1/2\lambda)K^2$ commutes with Ω because K commutes with Ω, and of course Ω commutes with Ω, so D commutes with Ω. Therefore we have

$$\Omega = \frac{1}{2\lambda}K^2 + D,$$

where D is a matrix that commutes with G, K, and Ω.

Let Q and P be the matrices that represent the particle's position coordinate and momentum and let m be the particle's mass. We assume Q and P are changed by space translations and Galilei transformations the way position coordinates and momenta always are. That implies the commutation relations

$$QK - KQ = i, \qquad PK - KP = 0,$$

$$QG - GQ = 0, \qquad PG - GP = -im.$$

We assume the system is just this one particle and all the matrices used to describe it are made from Q and P. The particle's position and velocity are the only variables. There is nothing to measure except where the particle is, which way it is moving, and how fast.

Then D is a matrix made from Q and P. We know that D commutes with K and G, that K commutes with all the matrices D can be made from except Q, and that G commutes with all the matrices D can be made from except P. It follows that D does not depend on Q or P. We can show this the same way we showed in the last chapter that V does not depend on R_1, R_2, R_3 or $\hbar K_1, \hbar K_2, \hbar K_3$. This means there are no matrices at all that D can be made from. It must be just the matrix 1 multiplied by a number.

Another argument to the same end is that if all the variables are made from the particle's position and velocity, then all the variables are changed by space translations or Galilei transformations, so the quantity represented by D, which is not changed by space translations or Galilei transformations, must be a fixed number like the particle's mass.

What we use from all this is that D commutes with Q. Another approach is to assume that G, K, and Ω are the basic building blocks and that all matrices are made from them. Then Q is made from G, K and Ω and, since D commutes with G, K, and Ω, it follows that D commutes with Q.

From this we get

$$Q\Omega - \Omega Q = Q\frac{1}{2\lambda}K^2 - \frac{1}{2\lambda}K^2Q = \frac{1}{2\lambda}(QK - KQ)K + \frac{1}{2\lambda}K(QK - KQ)$$

$$= \frac{1}{2\lambda}iK + \frac{1}{2\lambda}Ki = i\frac{1}{\lambda}K.$$

On the other hand, we know from Chapter 26 that

$$\left(1 + i\varepsilon\Omega - \tfrac{1}{2}\varepsilon^2\Omega^2\right)Q\left(1 - i\varepsilon\Omega - \tfrac{1}{2}\varepsilon^2\Omega^2\right) = Q + \varepsilon V,$$

where $V = (1/m)P$ is the matrix that represents the particle's velocity. This implies that

$$Q\Omega - \Omega Q = i\frac{1}{m}P.$$

Therefore

$$P = \frac{m}{\lambda}K$$

$$\Omega = \left(\frac{\lambda}{m}\right)\frac{1}{2m}P^2 + D$$

and

$$QP - PQ = \frac{m}{\lambda}(QK - KQ) = i\frac{m}{\lambda}.$$

Evidently λ is to be identified as $m/\hbar$ and $\hbar D$ is a constant that represents the value of the energy when the momentum is zero.

Here is the origin of Born's "strange equation." We have followed the story far enough to find the beginning.

PROBLEMS

28-1. Consider the matrices G_1, G_2, G_3 that we used for a single particle in Chapter 27. Show that these particular matrices commute with each other. Do the same for the matrices G_1, G_2, G_3 that we used for a system of two particles in Chapter 27.

28-2. Show that the commutation relations of G_1, G_2, G_3 with J_1, J_2, J_3 are satisfied by the matrices G_1, G_2, G_3 considered in the last problem with the matrices J_1, J_2, J_3 given in Chapter 24 for one particle, Problem 24-6 for one particle with spin, Chapter 26 for two particles, and Problem 27-7 for two particles with spin.

28-3. Work out the commutation relations of G_1, G_2, G_3 with K_1, K_2, K_3 for the matrices G_1, G_2, G_3 considered in the last two problems with the matrices K_1, K_2, K_3 given in Chapter 25 for one particle and Chapter 26 for two particles.

28-4. Let D be defined by

$$\Omega = \frac{1}{2\lambda}\left(K_1{}^2 + K_2{}^2 + K_3{}^2\right) + D.$$

Show the commutation relations of the ten matrices J_1, J_2, J_3; K_1, K_2, K_3; G_1, G_2, G_3; and Ω with each other imply D commutes with these ten matrices. The result of Problem 24-3 can be used here. Assume the system is just one particle. Let Q_1, Q_2, Q_3 and P_1, P_2, P_3 be the matrices that represent the particle's position coordinates and momentum. Assume they satisfy the commutation relations with K_1, K_2, K_3 and G_1, G_2, G_3 that correspond to the way position coordinates and momenta are changed by space translations and Galilei transformations. Suppose the particle has a spin and magnetic moment described by matrices Σ_1, Σ_2, Σ_3. Assume they are not changed by space translations and Galilei transformations so Σ_1, Σ_2, Σ_3 commute with K_1, K_2, K_3 and G_1, G_2, G_3. From the same assumptions and reasoning we used for one dimension, it follows that D cannot depend on Q_1, Q_2, Q_3 or P_1, P_2, P_3. Can D be a matrix made from Σ_1, Σ_2, Σ_3? Assume the spin and magnetic moment are vector quantities, so Σ_1, Σ_2, Σ_3 satisfy the commutation relations with J_1, J_2, J_3 that characterize rotations of a vector quantity. Then the only way to combine Σ_1, Σ_2, Σ_3 to make a matrix that commutes with J_1, J_2, J_3 is to use

$$\Sigma_1^2 + \Sigma_2^2 + \Sigma_3^2,$$

which is the matrix 1 multiplied by 3. Since D commutes with J_1, J_2, J_3, it must be a number multiplying the matrix 1. Assuming that, work out

$$Q_1\Omega - \Omega Q_1, \qquad Q_2\Omega - \Omega Q_2, \quad Q_3\Omega - \Omega Q_3$$

and show the results imply

$$P_1 = \frac{m}{\lambda}K_1, \qquad P_2 = \frac{m}{\lambda}K_2, \qquad P_3 = \frac{m}{\lambda}K_3.$$

This implies P_1, P_2, P_3 commute with each other. Show the commutation relations of Q_1, Q_2, Q_3 with P_1, P_2, P_3 are obtained when λ is identified as $m/\hbar$ and that Ω is then $(1/\hbar)H$ where

$$H = \frac{1}{2m}\left(P_1^2 + P_2^2 + P_3^2\right) + \hbar D.$$

28-5. Suppose the system is a single particle with no spin or magnetic moment. Let Q_1, Q_2, Q_3 and P_1, P_2, P_3 be the matrices that represent the particle's position coordinates and momentum. Let D_1, D_2, D_3 be defined by

$$G_1 = \lambda Q_1 + D_1, \qquad G_2 = \lambda Q_2 + D_2, \qquad G_3 = \lambda Q_3 + D_3.$$

Assume Q_1, Q_2, Q_3 and P_1, P_2, P_3 satisfy the commutation relations with K_1, K_2, K_3 and G_1, G_2, G_3 that correspond to the way position coordinates and momenta are changed by space translations and Galilei transformations. Show that these and the commutation relations of G_1, G_2, G_3 and K_1, K_2, K_3 with each other imply that D_1, D_2, D_3 commute with G_1, G_2, G_3 and K_1, K_2, K_3. From the same assumptions and reasoning we used for D, it follows that D_1, D_2, D_3 do not depend on Q_1, Q_2, Q_3 and P_1, P_2, P_3 and therefore must be numbers multiplying the matrix 1. This implies Q_1, Q_2, Q_3 commute with each other.

28-6. Consider a particle with a spin and magnetic moment, as described in Problem 28-4 and in Problems 24-6, 25-2, and 26-1. Let

$$G_1 = \frac{m}{\hbar}Q_1, \qquad G_2 = \frac{m}{\hbar}Q_2, \qquad G_3 = \frac{m}{\hbar}Q_3$$

as usual, and let

$$Q_1' = Q_1 + \delta\Sigma_1, \qquad Q_2' = Q_2 + \delta\Sigma_2, \qquad Q_3' = Q_3 + \delta\Sigma_3,$$

where δ is a real number. Show that Q_1', Q_2', Q_3' satisfy the same commutation relations as Q_1, Q_2, Q_3 do with all the matrices J_1, J_2, J_3; K_1, K_2, K_3; G_1, G_2, G_3; and Ω, but Q_1', Q_2', Q_3' do not commute with each other. This shows that for a particle with spin, the commutation relations of J_1, J_2, J_3; K_1, K_2, K_3; G_1, G_2, G_3; and Ω with each other and with the position matrices do not imply that the position matrices commute with each other.

REFERENCES

1. V. Bargmann, *Ann. Math.* **59**, 1 (1954).
2. V. Bargmann, *J. Math. Phys.* **5**, 862 (1964).
3. M. Hamermesh, *Group Theory and Its Applications to Physical Problems*. Addison-Wesley, Reading, Massachusetts, 1962, Chap. 12.
4. M. Hamermesh, *Ann. Phys.* **9**, 518 (1960).
5. J. M. Jauch, *Foundations of Quantum Mechanics*. Addison-Wesley, Reading, Massachusetts, 1968, Chaps. 9–10, 12–14.
6. T. F. Jordan, *Linear Operators for Quantum Mechanics*. Wiley, New York, 1969, Chap. 7.
7. E. P. Wigner, *Ann. Math.* **40**, 149 (1939).
8. E. P. Wigner, *Group Theory*. Academic Press, New York, 1959, particularly the Appendix to Chap. 20.
9. T. F. Jordan, *Am. J. Phys.* **43**, 1089 (1975).

INDEX